Design Tools for Engineering Teams

An Integrated Approach

Design Tools for Engineering Teams: An Integrated Approach
by Karen A. Schertz and Terry A. Whitney

Business Unit Director:
Alar Elken

Acquisitions Editor:
James DeVoe

Executive Editor:
Sandy Clark

Development Editor:
John Fisher

Book Editor:
Pamela Lamb

Editorial Assistant:
Jasmine Hartman

Executive Marketing Manager:
Maura Theriault

Marketing Coordinator:
Karen Smith

Production Manager:
Larry Main

Production Editor:
Stacy Masucci

Art/Design Coordinator:
Mary Beth Vought

Full Production Services:
Liz Kingslien
Lizart Digital Design
Tucson, AZ

COPYRIGHT © 2001 by Delmar, a division of Thomson Learning, Inc. Thomson Learning™ is a trademark used herein under license

Printed in the United States
1 2 3 4 5 XXX 05 04 03 02 01

For more information contact Delmar,
3 Columbia Circle, PO Box 15015,
Albany, NY 12212-5015.

Or find us on the World Wide Web at
http://www.delmar.com

ALL RIGHTS RESERVED. No part of this work covered by the copyright hereon may be reproduced or used in any form or by any means—graphic, electronic, or mechanical, including photocopying, recording, taping, Web distribution or information storage and retrieval systems—without written permission of the publisher.

For permission to use material from this text or product, contact us by
Tel   (800) 730-2214
Fax   (800) 730-2215
www.thomsonrights.com

Library of Congress
Cataloging-in-Publication Data

Schertz, Karen
Design tools for engineering teams /
Karen Schertz & Terry Whitney.
p. cm.
ISBN 0-7668-1227-8
Engineering design. I. Whitney, Terry. II. Title.

TA174 .S34 2001
620'.0042—dc21
2001027385

NOTICE TO THE READER

Publisher does not warrant or guarantee any of the products described herein or perform any independent analysis in connection with any of the product information contained herein. Publisher does not assume, and expressly disclaims, any obligation to obtain and include information other than that provided to it by the manufacturer.

The reader is expressly warned to consider and adopt all safety precautions that might be indicated by the activities herein and to avoid all potential hazards. By following the instructions contained herein, the reader willingly assumes all risks in connection with such instructions.

The Publisher makes no representation or warranties of any kind, including but not limited to, the warranties of fitness for particular purpose or merchantability, nor are any such representations implied with respect to the material set forth herein, and the publisher takes no responsibility with respect to such material. The publisher shall not be liable for any special, consequential, or exemplary damages resulting, in whole or part, from the readers' use of, or reliance upon, this material.

Design Tools for Engineering Teams

An Integrated Approach

Karen A. Schertz

Terry A. Whitney

DELMAR

THOMSON LEARNING

Australia • Canada • Mexico • Singapore • Spain • United Kingdom • United States

Preface

This textbook began as a series of conversations between the authors and their colleagues over five years ago. The conversations centered around the need to provide engineering students with a broader understanding of how all aspects of business affected the engineer, and the design process.

To succeed in today's business climate, the engineering professional must possess the understanding and ability to work successfully within the entire business organization. This includes the ability to make good decisions based upon current business objectives, access and interpret resources, communicate, problem solve, be creative and innovative, and function effectively in a group/team environment. These fundamental business themes, put into an engineering context, allow this book to have broad and timely appeal.

The latest trends in engineering education are inclined toward addressing the engineer's/designer's/technician's need to understand and apply their specific technical skills to the entire engineering and business process. The engineering education community has begun to emphasize a "holistic" approach toward the design process, and to recognize this as a crucial skill for tomorrow's engineering professionals.* The responsibilities of designers and technicians continue to overlap and interface with all aspects of the engineering process and the business organization. The understanding and application of this holistic approach is as important as the ability to perform a specific engineering task (such as operating a CAD application or interpreting a bill of materials). How any aspect of a design (cost, time, quality, market need, customer requirements, etc.) influences the overall business objective is fast becoming the concern of each member of the design profession.

The current state of engineering technology is one of rapid change and adaptation. Given the staggering pace of technological change in today's world, engineering needs to be considered from a broader vision. This textbook addresses the entire process of engineering design from concept to the delivered product — a whole systems approach. Rapidly shifting technology creates the need for people to be educated in the very real skills of critical thinking, problem solving, resource/information acquisition, and decision making. Engineers also need to be prepared to view change in a positive light and possess a fundamental knowledge of the overall process of engineering design. Technology requires judgments and decisions

---

* Refer to the current standards set by the three primary organizations in the U.S. for accrediting academic programs to meet specific criteria in the technical, engineering, and architectural disciplines: the Accreditation Board for Engineering and Technology (ABET), the National Architectural Accrediting Board (NAAB), and the National Association of Industrial Technology (NAIT).

beyond specific tasks and abilities; and it requires these decisions and judgments to be considered in the context of social, cultural, environmental, and ethical consequences. Students do need to possess specific technical skills such as, the operation of CAD programs, standard software applications, and knowledge of industry standards. But it is exactly because these specific skills are constantly changing that it becomes necessary for students to understand and have an awareness of the entire engineering process as it relates to the overall business plan.

This text is a must for anybody studying engineering/technology in the community college and university environment. The engineering design process is an activity that is common to all of the disciplines of engineering and technology education. These disciplines would include Architecture/Civil/Construction, Mechanical/Manufacturing, Electrical/Electronics, Transportation, Communications, or any segment of engineering that performs a design function. Most engineering/design textbooks are very specialized and do not provide the vision to see the design process in its entirety. The primary goal of this text is to provide today's technology students with the knowledge "tools" to integrate their technical skills into a successful and meaningful transition to the business world.

Organization of the Text

The general organization of this book stems from not only the authors' experiences as engineering/technology education teachers and professional designers, but from countless industry advisory councils and engineering professionals whose consistent advice is to educate students about the holistic process of engineering. In this vein, the textbook is organized from the general to the specific, while paralleling the engineering design process itself.

A common thread exists throughout the book: a case-study project integrates each chapter. This project considers all aspects of the business along the way (a reflection of today's workplace), and builds to an ultimate solution. The initial introductory chapters, Whole Systems Thinking, Teamwork, and Creativity/Innovation include a discussion of the premise and rationale for the book. Each subsequent chapter addresses one of the fundamental elements in the engineering design process. These fundamental elements are comprised of specific aspects of the design process including project management, design reviews, communication tools, problem-solving processes, and the engineering change process. Each of these topics is presented in the sequence it would commonly occur during the design process. This organizational theme is used as a model to simulate the realities of today's business environment.

The overall scheme of the book incorporates engineering industry design project scenarios, "industry interviews," and student activities/projects that provide immersion in the design process and contain the decision making requirements and consequences that all students will face, regardless of their specific roles in the engineering profession. The rationale being that unless a situation is perceived as "real" to students, they cannot feel a commitment to the process and be fully motivated and involved — the very ingredients that lead to true learning. Another of the components of this textbook provides the student with a metaphorical toolbox that contains "tools," or resources, to apply in various situations in both the educational arena and the workplace.

This text can be used either sequentially, if used as the primary text in a stand-alone course, or randomly, if used as a supplemental text that accompanies a specific engineering course. With the exception of the threaded "A-Team" case-study that acts as a unifying theme throughout the book, each chapter can be taught as a stand-alone topic. The textbook is structured to act as a launching point for the discussion of what role the engineering professional plays in the business organization, and how this manifests itself in the current business environment. The book should be used to address overarching fundamental business and engineering themes, and how they are integrated into the realities of the workplace. Specifically, the text is meant to provide students with an understanding and application of the engineering design process within the broader scope of today's business model. By using text design components such as the "Industry Scenarios," "Industry Interviews," and the "Toolbox" resources, students can integrate the content from any chapter into a discipline-specific project or function.

Preface

Features of the Text

This text has many special features that have been designed as teaching tools to provide learning experiences for students.

Activities and Projects — The assignments have been designed in a way that any engineering discipline will fit with the assignment. They are project-focused so as to allow students to use the text as a resource guide on how to proceed with their own projects. The activities are open-ended so instructors can customize them for their own instructional environment.

"A" Team Scenario — This feature provides the student with an example of how the concepts apply to real work situations. There are associated questions to each chapter scenario. The chapter scenario is tied to the concepts discussed in the chapter. The instructor may use these questions as discussion items or as homework assignments.

Toolbox — The Toolbox feature captures the specific tools required to accomplish the tasks presented in the chapter. It is a helpful tool for the instructor to aid in capturing critical information on the chapter material for test questions, report writing, or inclusion in a lecture.

Industry Interviews and Scenarios — This feature brings the real world into the classroom. The instructor may use this feature as a discussion item or reading and reflection assignment. Also they could be lead-ins or closings to a class lecture.

v

Design Tools for Engineering Teams

Sidelights and Chapter Quotes — These are meant to inspire reflective thoughts. The instructor can use these as thoughts for the week posted on the classroom board or for quick ideas to promote creative thinking with the class.

> The company that figures out how to harness the collective experience of its employees will blow the competition away.
>
> — Walter Wriston, former CEO, Citibank

Summary . . . Links to the Whole — The summary is designed to link the chapter concepts back to the whole picture of the integrated design process. It is the "why did we learn this?" answer.

Book Illustrations — The authors deliberately chose to use hand-sketch illustrations rather than slick, hi-tech photos. It is the authors' intent to convey that we still use our sketching talents to work through our design ideations.

Color Insert — This feature provides the student four real-world examples of the systems design process. Each example represents a different engineering field. These can be used when using the text or for classroom discussion.

Supplements

The Instructor's guide for this text includes, for each chapter of the text, the following materials.

- Suggested syllabus
- Term definitions
- Suggested Activities and Resources
- Transparencies created from PowerPoint® slides of key chapter points

Preface

About the Authors

Terry Whitney is currently a Training and Development Specialist in private industry, where he is involved in the design, development, and delivery of a broad range of learning initiatives that address human performance improvement. In addition to having spent twelve years teaching Engineering Design and Drafting at the high school and college level, he has over ten years experience in the engineering profession as a product designer in the computing and aerospace industries for companies such as Boeing, Honeywell, and Digital Equipment. Terry is currently a member of the American Society for Training and Development (ASTD) and the International Society for Performance Improvement (ISPI). He has previously authored an educational textbook on Commercial Architectural Applications. In addition, Terry was honored as a recipient of the 1994 Colorado Technical/Vocational School Faculty of the Year Award. The author holds undergraduate degrees in Business Information Systems from Arizona State University and Mechanical Design Technology from Glendale (Arizona) Community College.

Karen Schertz is a thirty-year veteran in the field of technical education. She has taught high school, community college, and at the university level in the areas of drafting, CAD, and design. For the past five years Karen was the Department Chairperson for the Advanced Technology Division at Front Range Community College in Fort Collins, Colorado. The author received the Faculty Member of the Year Award in 1990 from the Colorado State Board of Community Colleges. She previously published two drafting workbooks on basic drafting concepts and civil technology. Karen holds a Bachelor of Science degree in education from the State University of New York at Buffalo and a Master's Degree in Education from Penn State University. Presently Karen serves on the Rapid Product Development Alliance in Colorado and is a team recipient of an SME (Society of Manufacturing Engineers) grant award for Electronic Manufacturing program development.

Acknowledgments

The authors wish to express their sincere gratitude to the reviewers, industry experts, and production team that made this textbook a reality. The educational and industry experts who reviewed this textbook include:

- Tom Bledsaw, ITT Educational Services, Indianapolis, IN
- John Demel, Ohio State University, Columbus, OH
- William Dvonch, McHenry County College, Crystal Lake, IL
- Dorey Diab, Stark State College, Canton, OH
- James E. Folkestad, Colorado State University, Fort Collins, CO
- Jim Leonard, Colorado Manufacturing Competitiveness, Denver, CO
- Floyd Olsen, Utah Valley State College, Orem, UT
- James Overton, Southeast College of Technology, Memphis, TN
- Tom Singer, Sinclair Community College, Dayton, OH
- Kevin Standiford, ITT Technical Institute, Little Rock, AR

Industry Scenario contributors are:

- Brad Breckenridge, Student at Front Range Community College, Loveland, CO
- Caroline Fu, Learning Manager, The Boeing Company, Seattle, WA
- Reina De La Fuenta, PCB designer, Fort Collins, CO
- Bucky Wall, Program Management Training Manager, Microsoft, Inc., Redmond, WA
- Thomas Misage, CPIM, U.S. Small Business Administration, Fort Collins, CO
- Sean MacLeod, Business Development Manager, Stratos Product Development Group, Seattle, WA
- Richard E. Neal, President, IMTI, Inc., Oakridge, TN
- John Reynolds, PCB designer, Fort Collins, CO
- James Selle, Senior instructor, UCD Denver, CO

Mike Morison of Rocky Mountain Consultants, Longmont, Colorado contributed the civil project for the color insert. Don Lacombe and Maria Sheler-Edwards of Ford Motor Company were instrumental in providing art for the automotive project.

Preface

The authors would like to thank Delmar Thomson Learning for their belief and commitment to this project, in particular, Executive Editor Sandy Clark and Developmental Editor John Fisher. Sandy's initial belief in the concept, and John's tireless pursuit in making sure this textbook reflected the desires of the authors, while at the same time coordinating the technical aspects of publication proved invaluable. In the process of actually turning a rough manuscript into a professional publication, the authors would like to recognize the efforts of Liz Kingslien of Lizart Digital Design. As production designer, her organization and technical skills are unsurpassed, and her ability to keep the authors on schedule while accommodating their geographic, stylistic, and technical differences was truly amazing. Thanks also to the professional team of production experts including Mary Beth Vought, Art and Design Coordinator; Pamela Lamb, Book Editor; and Stacy Masucci, Production Editor. Special thanks goes to Victoria and Russell Inman for their technical assistance.

Terry Whitney would especially like to acknowledge his wife Janet for her patience, wisdom, and understanding; and their son Mark, who provides constant inspiration and clarifying perspective about what's really important in life.

Karen Schertz would like to acknowledge Gene Powell, who taught her to think out of the box; Anne Forman, who inspired her to become a teacher, and her husband, Leland, for his continuous belief and support of her dreams.

Karen and Terry wish to thank all the students who inspired this book. The book was written for them.

Design Tools for Engineering Teams

An Integrated Approach

Preface.. ii
    Organization of the Text iii
    Features of the Text v
    Supplements vi
    About the Authors vii

Acknowledgments viii

Chapter 1 | An Introduction to Whole Systems Thinking 1

    Introduction... 1
    Whole Systems and the Engineering Design Process 4
    The (Whole) System is the Solution 7
    Changing Reality of the Business Organization 10
    Engineering Design Models 12
    The Educational Community, Whole Systems
       Thinking, and the Engineering Design Process........ 16
    The Engineering Design Process...................... 18
    All Aspects of Design 23
    Summary . . . Links to the Whole..................... 24

Chapter 2 | Teams As a Tool in the Engineering
              Design Process....................................... 31

    Introduction... 31
    Orientation to the Concept of Teamwork.............. 32
    Types of Teams and Their Functions 35
    Development of Teams................................ 35
    Team Empowerment: How Is It Established
       and Maintained?.................................. 39
    The Cycles of Team Maturity 43
    Team Leadership and Team Control 45
    Summary . . . Links to the Whole..................... 47

Chapter 3 | Creativity and Innovation in Design................ 55

    Introduction... 55
    Qualities of the Creative Person.................... 58
    Principles of Creativity............................. 58

Table of Contents

    Methods for Developing Creativity 61
    Individual Creativity versus Group Creativity. 69
    Creativity versus Innovation 71
    Innovation through Failure 73
    Creativity, Innovation and the
       Engineering Design Process 74
    Innovation and the Use of Rapid Prototyping 76
    Summary . . . Links to the Whole. 78

Chapter 4 | Problem-Solving Processes for Design 83

    Introduction. 83
    Information Sharing 84
    Evolution of Problem Solving to Achieve
       a Quality Product 86
    The Problem-Solving Process 91
    Summary . . . Links to the Whole. 100

Chapter 5 | Communicating a Design: Making Your Case 113

    Introduction. 113
    Communication as a Strategic Planning Tool 114
    Communication Is Essential to Success,
       but Why Doesn't It Work? 115
    Making Team Communications Work 119
    Listening Skills Complete the Communication Loop 125
    Communication Techniques to Use with
       Clients, Customers, and Suppliers 126
    Negotiation Techniques 127
    Presenting Your Design: Making Your Case 128
    Summary . . . Links to the Whole. 130

Chapter 6 | Factors That Are Changing the
               Design Process 139

    Introduction. 139
    Better, Cheaper, and Faster. 140
    The Customer is at the Top of the
       Organizational Chart 142
    Virtual Teams. 144
    Virtual Design Process. 146
    Design for Manufacturability (DFM) 150
    CAD Software: 3D Modeling and
       What it Means for Design 150
    Product Data Management (PDM) 152
    Rapid Prototyping. 155
    How the Design Process Is Responding to Technology. . . 159
    How Recycling and Product Retirement
       Affect the Design 161
    Summary . . . or Links to the Whole. 161

Chapter 7 | From Concept to Delivery: Managing the Project 167

Introduction 167
History and Development of a Management System 167
The Old Organizational and Management Model 168
The New Emerging Organizational
  and Management Model 169
Components of Project Management 174
Planning the Project 176
Analyze Project Tasks and Flow 178
Management Software 183
HMMMM . . . There is Something Holding Up
  This Project 184
Improving Performance 187
Summary . . . Links to the Whole 190

Chapter 8 | Quality through the Design Review Process 199

Introduction 199
Components of Analysis and Design Review 201
Quality Standards and ISO 9000 204
Design Analysis Techniques 209
Peer Review Process 211
Customer Involvement in the Design
  and Review Process 213
Summary . . . Links to the Whole 213

Chapter 9 | Change: Learning to Love It 219

Introduction 219
The Nature of Change 221
Managing Change in the Engineering Design Process ... 226
Summary . . . Links to the Whole 230

Chapter 10 | Delivering the Product 235

Introduction 235
Commitments to the Customer 237
Key Elements of Project Completion 238
Summary . . . Links to the Whole 241

Bibliography 247

Index 249

An Introduction to Whole Systems Thinking

CHAPTER 1

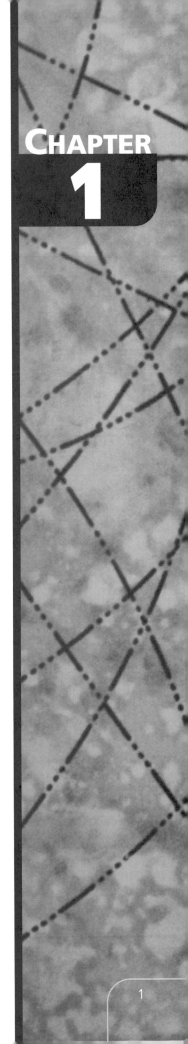

> "In physics, a linear system is, simply speaking, one in which the whole is equal to the sum of its parts (no more, no less), and in which the sum of a collection of causes produces a corresponding sum of effects. . . . The success of linear methods over the past three centuries has, however, tended to obscure the fact that real systems almost always turn out to be nonlinear at some level (the whole is now greater than the sum of its parts)."
>
> Paul Davies and John Gribben, *The Matter Myth*

Introduction

Well, what is a person to do? Everyone is telling us that this is the Information Age. Everything we need to know is literally at our fingertips. Information about any subject imaginable can be accessed with the click of a mouse. But what is often neglected is a discussion on what to do with all of this information. How does one determine which pieces of information are important to know, and which are not? Is the information correct, partially correct, or totally incorrect? Is any information missing that would improve the decision-making process? These are important questions to ask, because the information may be worthless, or even wrong, if it cannot be analyzed, put into the proper context, and seen as part of the big picture. The big picture, as defined in today's engineering environment, is the "whole system." How many times have you thought, "If I had only known an additional piece of information *(seen the big picture)*, I would have made a different decision."?

The need to apply a "holistic," or whole systems approach toward the design process is a critical skill for today's designers, technicians, and engineers as their responsibilities continue to overlap and interface with all aspects of the engineering and business process. Given the staggering pace of technological change in today's world, engineering, along with all aspects of business and life, needs to be considered from a holistic viewpoint.

Today, it is more important than ever before to develop the skills, ability, and knowledge to navigate the complexity of life, both personally and professionally. It is necessary to recognize the underlying dynamics that connect seemingly isolated events, actions, and decisions. Recent work in the field of industrial psychiatry tells us that there are specific core components that need to occur for success, regardless of the task or job involved. One of these components is the ability to recognize and appreciate the connectedness of all things and phenomena, or, *systems thinking*. It is an ability that cannot be overstated.

Without a whole picture of a situation or task, any decision or solution may simply lead to the existence of long chains of cause and effect interactions, mostly unintended, and usually undesirable. For example, consider a modern jetliner. It has a hydraulic system, a lighting system, a structural system, an air-handling system, a computer/avionics system, a plumbing system, and many other systems that make up the entire airplane. The airplane itself is a complex system, made up of a number of smaller systems, and it would be impossible to design one of these smaller systems and not impact the others in some respect. Without understanding how these specific systems affect one another, a field referred to as Systems Integration, there would be tremendous engineering problems when these individual systems are installed or operated as an integrated unit. Any number of unintended consequences could occur, such as physical part interferences, electrical power distribution overloads, or maintenance access problems, just to name a few. Let's call these unintended consequences "system effects." Without a whole systems approach to the design and production of an airplane, one system may not relate or "talk" to another, with the end result that, well, let's just say it certainly wouldn't be an airplane you would want to fly on! It is important to remember that system effects will be present whenever individual components are linked to a larger whole.

As a member of the engineering profession, whether that of technician, technologist, engineer, or architect, it is vital to possess abilities in areas such as problem definition, resource/information acquisition, design review, problem solving, technical communication, business systems, project management, and whole systems thinking. These qualities have become just as critical, and in some cases even more so, as the ability to operate a CAD software package, calculate a tolerance, or apply an engineering drawing specification. How one aspect of design, such as cost, time, quality, market need, or customer requirements, influences another is fast becoming the concern of every individual or design team member in a company. It's true, engineering and technical professionals do need to possess specific technical skills such as the ability to operate CAD software, use standard business software applications, and apply current industry standards. But because these specific skills are constantly changing, it has also become necessary to understand the entire engineering process, and the capability to *apply* this understanding in a manner that transcends the knowledge of a specific technical skill.

Today's employers assert that the most important knowledge a new employee can possess is not a specific technical skill, which can become obsolete as soon as the next software revision comes out, but the ability to understand how to work successfully within the overall system. These desired skills would include the ability to make quality decisions based upon current information, access resources, communicate, question, problem solve, and function effectively in a group/team environment. And that is what will be discussed in this book.

> The company that figures out how to harness the collective experience of its employees will blow the competition away.
>
> —Walter Wriston, former CEO, Citibank

Industry Scenario

— Caroline Fu, Manager, Quantum Shift Learning, The Boeing Company

As the creator of the Quantum Shift Learning organization within the Boeing Company, Caroline Fu is a primary catalyst for helping the company transform itself from a traditional compartmentalized enterprise into a systemic, whole systems, learning organization. With a background in physics, computer science, and mathematics, Caroline understands the intimate relationships among systems of all types and the leap that everyone of us must make in our thinking.

My role as manager of Quantum Shift Learning here at The Boeing Company is to facilitate the transformation of the Information Systems/Computing organization from a traditional linear-thinking approach to a true learning organization. We need to develop an attitude of "wholeness" in our thinking, to build people's capacity to think systemically. About three years ago, as a manager of Mechanical Engineering Systems, I was in the unenviable position of having to lay off employees during a down cycle in the aerospace business. In doing so, I realized that downsizing is analogous to an earthquake — it doesn't choose individual people, it chooses the weakest areas in the system. By that I mean that there were truly valuable employees being laid off because of rules and criteria (the weakest areas) that never addressed the best interests of the individual or the organization. As part of that experience, I developed a thesis for "turning the fear of downsizing into a constructive energy for positive change." The Quantum Shift Learning organization is a direct result of this belief. The company's goal is to transform its employees, and hence itself, from a mechanical, isolated Newtonian model to a systemic, whole systems, quantum thinking model. It is really more than just a model, it's a philosophy and a way of life. Whole Systems thinking is an attitude, a mental model that we can use in both our personal and professional lives.

Systems thinking is the cornerstone of how learning organizations think about their world, the essence of which lies in a shift of mind; seeing interrelationships rather than linear cause-effect chains, and seeing processes of change rather than just snapshots in time. Whole Systems thinking will allow us to improve our decision making, allow for more possibilities, reduce levels of fear, integrate and collaborate our problem-solving processes, and provide opportunities for growth. This is accomplished by promoting a dynamic tension between an ideal state and an actual state. In

Industry Scenario, continued

doing so, you can then continuously take intervening actions that will allow you to move closer to the ideal state. For example, recently there was an audit done to ascertain the level of preparedness for an upcoming computing systems benchmark. Of the five groups audited, the group best prepared had received "systems thinking" training. The audit determined that this group was the best prepared because they were more creative, responsive, systemic, and collaborative in their approach to the goal. Another example is a project that is underway to train our engineers to work with object-oriented CAD databases. We have found that most engineers tend to think procedurally (step-by-step), but with new object-oriented software applications, we need those engineers to think and problem solve in a more systems-oriented manner by integrating multiple points of view, diverse information, and systems philosophy.

By shifting our thinking to the "whole" and the interactions and relationships between its parts, and by developing "wholeness" as an attitude, we will be able to create a reality that will allow us to gain and apply a clearer understanding of our goals as both individuals and as a company.

Whole Systems and the Engineering Design Process

> The obstacle to knowledge is not ignorance, but the illusion of knowledge.
> —Daniel J. Boorstin

The **engineering design process** is the cornerstone of the engineering profession. It is the process used to accomplish the art and science of engineering. Engineering has been described as the practical application of the knowledge gained from the pure sciences, or, to describe it another way, of turning imagination into reality. As a future member of the engineering profession, it will be your goal to solve problems of pragmatic concern, in a way that is both practical (function) and elegant (form). This goal is usually accomplished by incorporating the engineering design process and the creative design process.

The engineering design process is often described as an organized series of actions that a designer takes when solving any type of engineering problem. But first and foremost, the engineering design process is a decision-making process. Since every problem is unique, the actions taken will sometimes vary. It is important to consider each action in a given process to determine if it is necessary. However, no matter what the quality of the final solution, there will always be room for improvement. Due to this fact, the engineering design process must be looked upon, not as a linear, step-by-step, start-to-finish process, but as pieces in a puzzle (see **Figure 1.1**), which when put together properly form a complete and integrated solution. And although this puzzle is put together in a specific sequence, it is important to realize that the completion of each puzzle piece may create the need for reexamination of other pieces.

Chapter 1 | An Introduction to Whole Systems Thinking

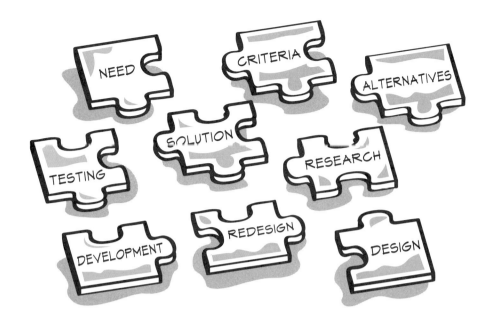

Figure 1.1
The Engineering Design Process — Pieces of a puzzle

Traditionally, designers were trained in design fundamentals, the nature of materials, the capabilities of tools and equipment, and manufacturing/construction processes, to become proficient in their field. Now in addition to these skills, a true understanding of the customers' needs and an awareness of all aspects of the business is required to allow the designer to develop the most optimal solutions, in terms of cost, quality, time, and specific other criteria. Take a moment and think about virtually any object around you. It was designed to meet specific criteria and specifications. For instance, pick up an ordinary writing pen and study it for a moment. Think about all of the factors that were considered by the designer when designing the pen:

- What material should it be made of?
- What shape should it be (round, hexagonal, triangular, or tapered)?
- What color(s) should it be?
- Clip or no clip?
- Eraser or no eraser?
- Push-button or cap?
- Disposable or refillable?
- Structural integrity (What happens when it is dropped and accidentally stepped on?)
- Weight

But these are just the technical considerations. It is crucial that the designer understand all other factors as well to design the optimal product. Good design anticipates and incorporates these related factors into the design, safety, cost, market trends, environmental impacts (what happens to it when it is used improperly or thrown away?), packaging, distribution, and storage. This is an illustration of a systems approach to design. The need for close integration of all aspects of the product cannot be emphasized enough. This whole systems approach is a process that fulfills two fundamental purposes: (1) It makes sure that we ask and understand the right questions before designing the answer, and (2) It coordinates, focuses, and balances all technical and related efforts in the design and development process.

The whole systems approach is invaluable for satisfying the main driver of today's business environment — competitiveness. The issue of competitiveness has arisen from the incredible availability of similar products and services found in the world today. It is a natural consequence of the global economy and technological advances in production, distribution, and availability of information and products through sources such as the Internet.

What may at first seem like peripheral subjects to the design process are actually critical issues that must be considered in the engineering design process. The following issues are just a few of what necessitate the need for a whole systems design approach: Federal and State regulations, environmental impacts, consumer perceptions, economic/societal needs, cultural trends, and demographic shifts. A varied list with complex dependencies and interrelationships exists that must be addressed in the engineering design process.

It is now recognized that the ability to create, manage information, see complex relationships and patterns, and learn continuously will replace capital, labor, and natural resources as the most desirable assets not only for business and industry, but in society as well. Two major forces, rapid technological advances and global competition, have profoundly changed, and will continue to change, the fundamental nature of work. This has necessitated the creation of a new business paradigm in which intellectual capital has become more important than physical/financial capital — the knowledge-based workforce.

> The enlightened organization requires that everyone continually challenge prevailing thinking, and think systematically by seeing the big picture and balance short- and long-term consequences and decisions.
>
> —Peter Senge

CHANGING BUSINESS PARADIGM

INDUSTRIAL WORKER (YESTERDAY)	KNOWLEDGE WORKER (TODAY)
Physical work	Intellectual work
Factor of production	Knowledge producer
Question nothing	Question everything
Do as you are told	Determine what to do
Repetitive tasks	Continuous challenges
Segmented work	Holistic work
Direct supervision	Autonomy
"A strong back"	"A strong resume"

As the complexity of technology advances, so will its interdependence, requiring individuals with skills in designing and managing projects from a holistic viewpoint. In turn, this has created a new business structure, a transition from that of a hierarchical organization to that of a learning organization (see Figure 1.2).

Figure 1.2
Transition to a learning organization

The (Whole) System is the Solution

Webster's Tenth Collegiate Dictionary defines the word *holistic* as: "Related to or concerned with wholes or with complete systems rather than with the analysis of, treatment of, or dissection into parts; *holistic* ecology views man and the environment as a single system." *Webster's* also defines *system* as "a regularly interacting or interdependent group of items forming a unified whole; a group of interacting bodies under the influence of related forces."

Consequently, a definition of whole systems thinking would be "the ability to conceptualize problems in a manner that accounts for the entire structure or system." This leads to a fundamentally sound solution, utilizing the interconnectivity of all things and the corresponding effect one action has on the system as a whole.

A whole system can be seen as dependent upon all of its parts, all of which contribute to its overall state of harmony. Likewise, the parts are seen to be dependent upon the harmony of the whole system. It is also important to remember that the parts of a whole system can be seen as holistic entities in their own right, and that a whole system may also be viewed as part of a larger system. To illustrate this point

> Pit a good performer against a bad system, and the system will win every time.
>
> — Geary A. Rummler

more clearly, remember the earlier scenario of a jet airliner. The hydraulic system would be representative of a whole system within a larger system — the airplane. To take it a step further, the airplane could also be viewed as a whole system within a larger system defined as air transportation, as illustrated in **Figure 1.3**.

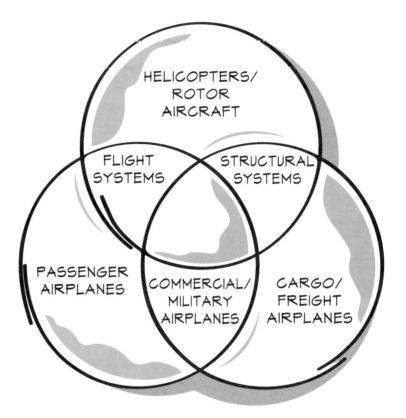

Figure 1.3 Whole system model of air transportation system

From an engineering design standpoint, it is crucial to consider these system relationships. It is also very important to remember that the more complex a system is, and the more interactions that take place between the system's components, the more difficult it is to predict the full effects of any given action.

Whole systems thinking identifies and addresses long-term and structural problems. It recognizes the interdependency between you, your customers (customers being anyone who depends on you), and your suppliers (suppliers being anyone you depend on). Whole systems thinking also recognizes that seemingly innocuous changes at any point in the system can have wide-ranging unintended impacts on the entire system.

Historically, science and technology have utilized a reductionist approach to problem solving. By reductionist, we mean the theory of reducing a problem to its simplest and most definable level. In doing so, we have, in many cases, failed to recognize the effects of actions outside of their immediate context, or in other words, we have failed at developing solutions within a holistic framework of under-

Chapter 1 | An Introduction to Whole Systems Thinking

> **Industry Scenario**
>
> Soon to go into effect is the legal requirement that children under the age of two will be required to be placed in a separate seat for airline flights. The requirement is an attempt to provide increased safety for infants, who can currently travel on the laps of an adult during airline flights. Although well-intentioned, there are some critics of this legislation that have taken a whole systems approach to evaluating this solution. Specifically, the critics have proposed that because a separate seat is now going to be required, which causes an additional ticket to be purchased, many families will choose to drive to their destination instead of flying because of the additional cost. Since many more people are killed in automobile accidents than air travel, the goal of improved infant safety could actually take a step backwards. What are your thoughts?

standing, a systems viewpoint. In the past thirty years technology has developed so fast, and there is so much, that traditionally organizations have divided up jobs into little pieces and had individuals focus on those pieces. The problem with this approach is that people never had the opportunity to see where their piece of the puzzle fits into the big picture. Consequently, when the pieces were put together, there tended to be unforeseen side effects and problems.

> **Industry Scenario**
>
> As a result of technological advancements in a variety of areas, home security systems have become more inexpensive. These low-cost, home security systems have become widespread, and because of the technology they employ, one of their characteristics is that they are easily activated, which is a presumably positive characteristic. However, because such factors as human error and varying levels of activation have not been included into the overall equation of most home security systems, there have been very real negative consequences. For example, the Philadelphia Police Department reported that over a three-year period, only 3,000 out of a total of 157,000 calls from automatic home security systems were actual attempted break-ins, and fully half of those calls were caused by user errors. As a consequence, the full-time equivalent of fifty-eight officers were diverted to answer these useless calls, allowing crime to go unchecked elsewhere. This illustrates an example of the unintended consequences that can occur when all aspects of a system are not addressed.
>
> *(Source:* Philadelphia Police Department, 1996)

Changing Reality of the Business Organization

Developments in new technology, namely computing software and hardware, now allow for information to be disseminated easily throughout a company. This has led to a more knowledgeable workforce, greater complexity, and the increased availability of information. Many organizations that are now failing are those that placed a strong emphasis on centralized management with strict command and control structures. This type of corporate structure is rarely desirable, or even possible, in today's business organization. For example, the information that today's engineering organizations generate is now available to many other departments within a company, and in some cases, even customers. These other departments may use the information from engineering for purposes such as inventory planning, production scheduling, purchasing, labor forecasting, product delivery schedules, and customer service. Until recently, engineering organizations, and every other organization within a company, tended to be isolated islands of information. And there weren't many bridges that connected these organizational islands, as shown in **Figure 1.4**.

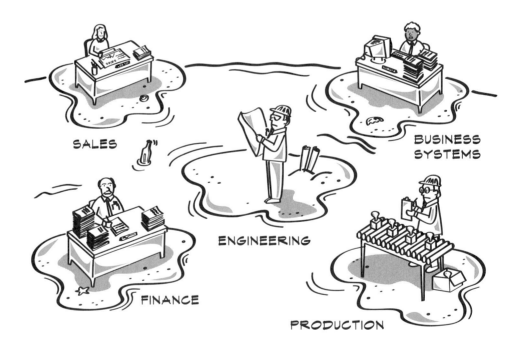

Figure 1.4
Isolated organizational islands

Today, with the advent of advanced technologies like Web-based computing servers, virtually every function (engineering, production, sales, finance, and human resources) within a business is connected with immediate and current information. This has created the opportunity and necessity to better understand how and why this information is used. The engineering community is now being asked to consider an array of questions such as: What does the project cost? How are those costs being measured for success? Who exactly are my customers? What does the customer really want? What is involved in project integration?

These questions and many more require a breadth of knowledge about business, economics, finance, and marketing that were not necessary in yesterday's world. One way business is attempting to provide an entire array of information to its employees is through the use of **knowledge management** systems. Knowledge management attempts to capture the collective wisdom of a business and provide it to the individual on demand, since we now know that decisions that make or break a company are not those made by strategists at the top. Whether in war or in commerce, it's the sum total of countless decisions made every day on the front lines that determine the course of future events. Their amassed weight can create a momentum, or a chaos, far beyond the power of senior leaders to redirect.

Therefore, success, in business or other large-scale endeavors, depends on good individual, daily decisions outweighing bad ones over time. And the most important thing top management can do to ensure success is to empower people throughout the organization to make good decisions. Partly, this is a question of simply granting the authority for decision making, and establishing accountability for decisions made. But much more importantly, it's a question of equipping people with the *knowledge* required to make decisions well.

When bad decisions are made, after all, it is rarely because the knowledge did not exist in the organization to make them well. It's because this knowledge wasn't brought to bear on the decision, because it was buried in a corner of the organization where it could not be found in time, or because it was never sought in the first place. Applying the fullness of an organization's knowledge to its decisions means working hard to represent it, transfer it, make it accessible, and encourage its use.

Today's markets are driven by cost, trends, and customer service as much as by quality and technology. Because of these realities, engineers, designers, drafters and technologists now need to be aware of how their designs fit into the broader function of product development. In turn, they need to understand how this function fits into the entire product development cycle. And finally, this knowledge needs to be carried one step further into the areas of sales and marketing to understand exactly what the company is trying to sell and produce, to whom, and what for. Hence the need for an understanding of a whole systems philosophy.

> Technology has made it so easy to amass information that it can get in the way of acquiring knowledge that has strategic value for a company.
>
> —Ledet

Industry Scenario

The Story of Southwest Airlines

Southwest Airlines is a terrific example of a company that makes sure its employees understand the Whole System philosophy. Southwest is annually rated at the top in the categories of on-time reliability, customer satisfaction, and low ticket costs. Southwest's philosophy can be tied directly to its defined purpose: the ability to move large numbers of people efficiently, friendly, and cost-effectively, from one destination to another, or as Southwest Airlines puts it, "faster, cheaper, and more fun." So when any employee within one of the company's organizational systems (such as customer service, airplane acquisition and maintenance, computer information systems, or marketing) needs to make a decision, they ask themselves if that decision fits into the company purpose, or the Whole System. The reason that Southwest doesn't offer food on its flights (it is cheaper and makes it quicker to turn around flights), does not assign seats on its flights (reduces overhead costs), flies only one kind of aircraft (lower maintenance costs), and emphasizes customer service (makes flying more pleasant and fun) is directly linked to the company purpose. Every decision that is made, is made only after asking, Does this help achieve the goal of moving large numbers of people efficiently, friendly, and cost-effectively from one destination to another? Based upon Southwest Airlines spectacular growth and rankings in a competitive industry, the answer has been a definite yes.

Engineering Design Models

There are a number of related approaches to engineering design that encompass the fundamental principle of whole systems thinking and our interdependence on people, things, and processes. These related engineering design models all encompass a basic premise that has been true for the span of human existence: you need other people to know things. This sharing of knowledge, varying viewpoints, and ideas leads to better, faster, and more efficient problem solving. Accumulated knowledge and the availability of information in the engineering profession is just as explosive as it is in the rest of society. The ability and tools to gather and archive information is available to more people, more quickly, and with less effort than ever before. Although this may appear to make individuals less dependent on others, paradoxically, it increases our reliance upon others. This is because there is simply too much information for one person to process, analyze, and manage. Because of this, the use of cross-functional teams and whole systems thinking has become a natural evolutionary step in the engineering world, the work world, and society itself. Let's look at some of the systems approaches used by the engineering world to address the constantly evolving engineering design process.

Chapter 1 | An Introduction to Whole Systems Thinking

Concurrent Engineering and Whole Systems Design

Concurrent engineering is getting the right people together at the right time to identify and resolve design problems. Concurrent engineering is designing for assembly, availability, cost, customer satisfaction, maintainability, manageability, producibility, operability, performance, quality, risk, safety, schedule, social acceptability, and all other attributes of the product. As an integrated approach to design, production, and customer service, concurrent engineering emphasizes the advantages of simultaneous, or concurrent, product design by utilizing individuals from various areas of the business in the up-front concept and design phase, with special emphasis on customers and their needs. Concurrent engineering promotes an integrated, holistic approach to the design effort, and by focusing on all aspects of design simultaneously, errors can be detected earlier in the process, preventing more costly and time-consuming rework in the production and testing phases. This process involves the integration of people, processes, standards, tools, and methods to achieve a faster and, more importantly, a better solution to the design problem (see **Figure 1.5**).

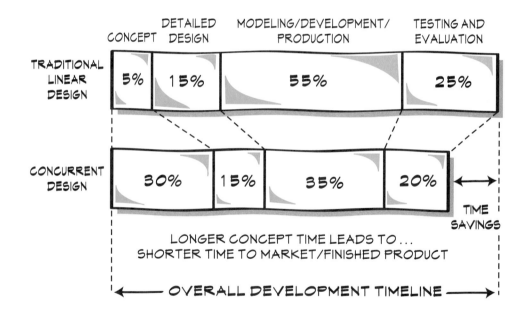

Figure 1.5 Concurrent engineering process flow

Life Cycle Engineering

The concept behind **life cycle engineering** is that the entire life of a product should be evaluated at the beginning of the design process. The optimal design cannot be achieved unless performance costs (both initial and ongoing), reliability, maintainability, and market trends are addressed first. Good design will also consider product replacement timelines, and the eventual withdrawal or disposal of the product from the market, and what product will replace it. These include issues such as software upgrades, technology advancements, changing customer needs, market trends, and cultural shifts.

Integrated Product Development (IPD)

Integrated product development (IPD) is the process of developing a product such that all sub-processes required for its development are integrated seamlessly. More simply stated, a product should be designed and developed to satisfy all of the conditions the product will encounter, not just the initial ones. For example, it is not enough that a computer is affordable and performs well. An integrated product development effort would also consider the computer's appearance (look at the Apple® iMac™ computer!), its ease of installation and/or assembly, ease of maintenance, service or repair, and its method of disposal (trade-in, sell, or throw away). IPD would evaluate the market segment the computer is designed for, the desired design/market life of the product, and the requirements for meeting United States. and international standards. Then there are the production considerations: Is the product easy to obtain/build parts for, assemble, ship, package, and distribute? By evaluating all of these factors in the initial design, there is a much greater likelihood that unforeseen circumstances will not accumulate and damage the product's chances in the marketplace.

Knowledge-Based Engineering (KBE)

Knowledge-based engineering (KBE) is the development of computer models to simulate the best-known engineering processes. Typically, KBE is used by creating a set of engineering data. This data is comprised of components such as CAD data, manufacturing data, tooling data, and structural information. And this data is then used in an integrated manner to develop a more detailed and comprehensive design plan, while also allowing for better analysis and evaluation throughout the process.

Total Quality Management (TQM)

There are several definitions for the philosophy known as **total quality management (TQM),** but at its heart is a set of management practices designed to continuously improve the performance of the entire business process to achieve customer satisfaction. TQM calls for the integration of all organizational activities to achieve the goal of serving customers. It seeks to achieve this goal by establishing process standards, maximizing production efficiency, implementing quality improvement processes, and employing integrated teams to more effectively design customer-driven products. Total quality management is a philosophy, a set of tools, and a process. Specifically, a total quality management approach differentiates cost versus price, emphasizes defect-free products, strives for the elimination of all nonvalue-added activity in the production process, and delivers competitive priced products timely and efficiently to the marketplace. Above all, TQM strives for satisfaction of all of its customers, both internal and external.

Chapter 1 | An Introduction to Whole Systems Thinking

Causal Loop Diagrams and System Dynamics

From both a professional and personal standpoint, a basic knowledge of systems behavior is important. One approach to systems thinking is with the use of causal loop diagrams and system dynamics. System dynamics is a methodology for understanding that everything around us, and including us, is an interrelated system. Although not all actions, thoughts, or events have equal effect, all of these can result in surprising causal consequences. A major component of system dynamics is the causal loop diagram. A causal loop diagram is a visual tool designed to allow quick analysis of a cause and effect situation. In other words, by using this tool, a person can simulate, then readily identify the effects of a proposed solution, in its simplest terms. The causal loop diagramming process begins with the relationship that exists between any pair of variables. For example, a bank account balance and the amount of interest paid on a bank account balance are directly related to each other. These variables produce a link, in that when one variable is changed, it has an impact on the other variable. This impact can be either a desired outcome (positive feedback) or an undesired outcome (negative feedback), as shown in Figure 1.6. An example of a simple causal loop diagram using the bank account scenario follows:

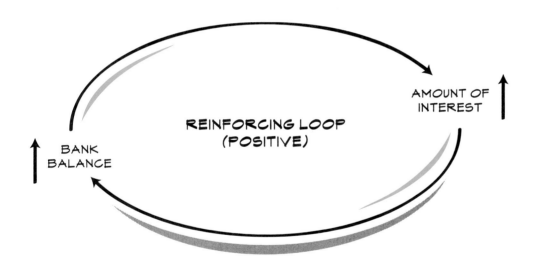

Figure 1.6
Causal loop diagram

Where as the bank balance increases, the amount of interest paid increases, thus causing the bank balance to increase. And so it goes, in a perpetual, positive-feedback, reinforcing loop.

The Educational Community, Whole Systems Thinking, and the Engineering Design Process

As business and industry continue their rapid pace toward a greater emphasis on a whole systems approach toward design, development, operation, and management, the educational community has also embraced the whole systems concept in their academic offerings and requirements. Three of the primary organizations in the United States for accrediting academic programs to meet specific criteria in the technical, engineering, and architectural disciplines are the Accreditation Board for Engineering and Technology (ABET); the National Architectural Accrediting Board (NAAB); and the National Association of Industrial Technology (NAIT). Each of these organizations has established requirements for their areas of expertise that ensure that students who graduate from accredited programs will be prepared to enter the profession.

The Accreditation Board for Engineering and Technology (ABET) is recognized in the United States as the sole agency for accreditation of educational programs leading to degrees in engineering. The current basic level accreditation criteria regarding program outcomes and assessment requires that engineering programs must demonstrate that their graduates have, in addition to fundamental mathematics, science, and engineering, knowledge and skills related to the their chosen discipline:

- An ability to function on multidisciplinary teams
- An ability to communicate effectively
- The broad education necessary to understand the impact of engineering solutions in a global and societal context
- A recognition of the need for, and an ability to engage in lifelong learning
- An ability to design a system, component, or process to meet desired needs
- An understanding of professional and ethical responsibility
- A knowledge of contemporary issues

The professional architectural education programs in the United States have a similar process. The National Architectural Accrediting Board (NAAB) is the accreditation board who sets criteria for educational institutions that offer professional architectural programs. The NAAB has established student performance criteria that a program must meet to ensure that its graduates possess the skills and knowledge necessary to pursue professional licensing. The list of performance criteria begins with fundamental skills and knowledge, continues with technical skills and knowledge, and concludes with a focus on practice and societal roles. This sequence is intended to foster an integrated approach to learning that cuts across subject categories. For purposes of accreditation, graduating students must demonstrate awareness, understanding, or ability in the following areas:

Chapter 1 | An Introduction to Whole Systems Thinking

- Research skills
- Critical thinking skills (comprehensive analysis and evaluation)
- Collaborative skills (working in teams)
- Human behavior (relationships between human behavior and the physical environment)
- Human diversity (awareness of the diversity of needs, values, and social patterns, and their implications for society and the role of the architect)
- Western traditions (understanding of Western architectural traditions in architecture, landscape, and urban design, as well as the climactic, technological, socioeconomic, and other cultural factors that have shaped them)
- Non-Western traditions
- National and regional traditions
- Environmental conservation
- Environmental systems
- Life-safety systems
- Building Service Systems
- Building systems integration
- Legal responsibilities
- Building economics and cost control
- Comprehensive design
- Program preparation (includes needs assessment, review of site conditions, relevant laws and standards, space and equipment requirements, etc.)
- Practice organization and management (includes office organization business planning, marketing, financial management, and leadership)
- Ethics and professional judgment

These criteria are in addition to, and on the same level as, all of the technical and theoretical knowledge that architectural students must learn. It is easy to see that a whole systems approach is taken by not only business and industry, but by the educational community itself, in all engineering disciplines.

In the field of Industrial Technology, which focuses on the management, operation, and maintenance of complex technical systems, the National Association of Industrial Technology (NAIT) is the recognized professional association for accreditation of industrial technology programs in colleges, universities, and technical institutes. Additionally, NAIT serves to provide professional certification of industrial technologists in pursuit of their continued professional development.

As defined by NAIT, Industrial Technology is a field of study designed to prepare technical and/or technical management-oriented professionals for employment in business, industry, and government. Industrial technology degree programs and professionals will be involved with the following:

- The application of theories, concepts, and principles found in the humanities and the social and behavioral sciences, including a thorough grounding in communication skills
- The understanding of the theories and the ability to apply the principles and concepts of mathematics and science and the application of computer fundamentals
- The application of concepts derived from, and current skills developed in, a variety of technical and related disciplines, which may include, but is not limited to, materials and production processes, industrial management and human relations, marketing, communications, electronics, and graphics
- The completion of a field of specialization, for example, electronic data processing, computer-aided design, computer integrated manufacturing, manufacturing, construction, energy, polymers, printing, safety, or transportation

Note that the first of the four major requirements refers to acquiring the skills and abilities necessary for applying a whole systems approach to technology.

The Engineering Design Process

Engineering design process is a sequential series of steps or actions a designer takes when solving a problem or creating a new design. Because every engineering problem or design has its own unique set of concerns and criteria, it is impossible to describe a definitive process that would be used by every designer in every instance. There is, however, a generally accepted methodology used in the design profession, which can be effectively applied to most design projects. Although it is commonly broken down into 5, 6, 7, or more individual steps with an assortment of names to identify each of these steps, the engineering design process can also be looked at as a series of questions to be answered:

THE ENGINEERING DESIGN PROCESS

DESIGN PROCESS STEPS	QUESTION TO BE ANSWERED
Identify the need	Do our customers have a need, and if so, what is it?
Define the criteria for the need	What does our team's solution need to do?
Exploration/research/investigation	How do we solve the design problem?
Generate alternative solutions	What are the possibilities?
Choose a solution	What is our best possible solution?
Detailed design	What are the details of our design?
Modeling/development/production	Is it turning out the way we thought it would?
Design testing and evaluation	Did the design do what our customers wanted?
Redesign and improvement	How can we do it better next time?

It is very important to realize that each of the steps are of equal importance. Short-changing or omitting even one of the steps will have serious negative impacts on the final design solution. Remember . . . whole systems thinking! In addition, the designer must be aware that it is very possible to be moving in either direction in the design process, based upon a constant evaluation of how the design is evolving. The engineering design process is one of continuous improvement and nonlinearity, meaning that sometimes it is necessary to back up a step or two before moving forward again. Let's take a look at each of the questions to be answered, and their associated steps, in the engineering design process (see **Figure 1.7**).

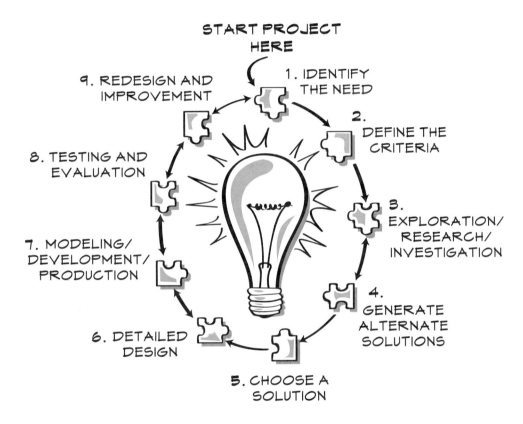

Figure 1.7
The Engineering Process

1. **Do our customers have a need? Identify the need**

   The first step in any engineering design process is to ask, Do I have a need? And if so, what exactly is it? In many cases, this obvious first step is not really evaluated. It is assumed to be known, and therefore, it usually means different things to different people, and the end result is that a product is defined and built that does not meet the exact expectations of the customer or the company. The world is filled with quality products that nobody needs or wants. Therefore, it is vital to specifically define what the need is and why.

2. **What does our team's solution need to do? Define the criteria for the need**

   The next question to answer is the determination and selection of the product's performance criteria. By this, we mean that specific criteria should be developed that explains exactly what the product should be able to do, how well it should be able to do it, and at what cost. The criteria should be developed as a collaborative effort with the customer if at all possible. This set of performance criteria, or specifications, should include basic information about cost, weight, size, durability, maintainability, and other characteristics dependent upon the function of the product.

3. **How do we solve the design problem? Exploration/research/investigation**

    In asking how to solve the design problem, it is important to first ask a number of *other* questions. The importance of asking questions, and asking them often, cannot be overemphasized. Questioning the project's progress on a regular basis will make sure that the design process is still going down the road to your stated goal, and not taking a detour and arriving at an unintended destination. Questions will help you sort out all the information you need to gather in this piece of the engineering design process puzzle. This information will come from many and varied sources, such as customer research, competitive analysis, current products, past products, product literature, peers, field experts/consultants, professional organizations and institutions, and many others. This information will give you an idea of what's out there, but just as importantly, what's not out there.

4. **What are the possibilities? Generate alternative solutions**

    Now that you have defined your product criteria and analyzed these gathered stacks of material related to your problem, it is important to generate a number of alternative solutions based upon this information. In doing so, it is good to generate various ideas without making immediate judgments as to their value. Although they may initially appear infeasible, by doing so, you will often end up incorporating some aspects of these designs into your final solution. In addition, these "wild" ideas may stimulate other ideas that would have otherwise gone unnoticed. Creativity and nonjudgment are the key ingredients to generating a number of alternative solutions. In the end, it is desirable to have a number of practical and not-so-practical solutions to your design problem. This is how unique and innovative solutions come about.

5. **What is our best possible solution? Choose a solution**

    Choosing a solution among all of the possible alternatives is the most challenging component of the engineering design process. It is at this stage that a decision must be made as to what appears to be the optimal design solution. Part of determining the best possible solution to the design problem is evaluating various trade-offs and the associated benefits and drawbacks that exist with each potential design. This will normally involve a team of individuals who possess expertise in a given aspect of the business, such as marketing, production, customer support, business systems, and others. It is the project team's responsibility to consider all of the various aspects of the business, weigh the positive and negative impacts to each, resolve any potential conflicts, and select a solution.

6. **What are the details of our design? Detailed design**

    At the detailed design stage of the engineering design process, it is necessary to document all of the design aspects, specifications, materials, procedures, and processes required to produce your product, be it an office building, a software application, or a toothbrush. Regular design review meetings are held to discuss how all of the various components of the design will be accomplished in the most seamless way possible. These design review meetings involve close teamwork among all of the involved groups, such as engineering, production, tooling, and finance.

7. **Is it turning out the way we thought it would? Modeling/development/production**

    As the product begins to take shape in the development and production/construction process, both literally and figuratively, the design teams should be making sure that the product is taking on the original design vision and will be able to meet the performance criteria, specifications, and goals established in the initial phase of the design process. This is accomplished through the use of detailed models, prototypes, simulations, and limited production of the product. Involving the customer at this stage to make sure that everyone is still in alignment with regards to the final outcome is critical to the project's success.

8. **Did the design do what our customers wanted it to do? Design testing and evaluation**

    The product has now become a reality. It is time to perform extensive testing and evaluation to see if the product will truly meet goals and expectations. The widespread use of testing teams has become a critical step in the design and development process, because it helps to ensure product reliability prior to the sale of a product, thus increasing customer satisfaction and reducing customer support costs.

9. **How can we do it better next time? Redesign and improvement**

    Because one of the factors in any design project is time, often products are developed without all of the optimal characteristics. Along with other factors like advances in technology and changes in customer desires, products are always being improved. Just as importantly, product/project development processes are being improved as well. Therefore the engineering design process is never complete. The continual quest for improvement and innovation is the motivation that has driven designers throughout history.

Industry Scenario

— *Andy Sweet, Software Engineering Manager, Qpass, Inc.*

As a software engineering manager for a growing, energetic, cutting-edge technology company, Andy is intimately involved in the entire business and engineering process for his company. Qpass Inc. designs, develops, and markets a product that allows the purchasing of digital content, such as reports, surveys, market research, and various other types of documents, over the Internet. The company views what they are providing to the customer as a service as much as a product, with the goal being to deliver a "frictionless" purchasing experience. In other words, the service should be as simple as possible for the customer to use. As a manager of a design team, Andy stresses the importance of the truly integrated engineering design process.

My responsibility as a software engineering manager is to oversee and manage the design and development of a product, with the constant reminder to myself about the importance of providing a truly user-friendly experience to our customers. In the case of Qpass, we actually have more than one customer base. The Internet content provider who wishes to sell digital content and the individuals who utilize the service are both customers of Qpass. In designing and developing the product, we realize that even if the product is terrific, it is of no value to the customers if it cannot be used easily. So, in addition to the development effort, up-front testing and customer support are huge issues. Since we base the service on a large volume of transactions with a small profit margin, our profits go down rapidly if our customers need to call customer service very often. So it becomes crucial that our service is reliable and easy to use.

In order to produce the best possible product and service, we make extensive use of teams in a totally integrated effort. Our engineering design process begins with the functional specifications being given to a product management design group leader, who generates a conceptual design framework. The conceptual design is then reviewed by representatives of the engineering team, the project management team, the marketing team, the customer service team, and the test and evaluation team. After addressing the concerns, suggestions, and ideas from these representatives, we begin implementation of a detailed product design. We also must address continual changes requested by the customers, since their requirements are driven by a rapidly changing technology and marketplace. After completion of the product, it is sent for testing and evaluation to our internal test team, where it must pass a significant number of test cases and scenarios before moving to the next step in the process. After successfully completing the test and evaluation stage, final changes can be incorporated into the product.

Because the software industry is relatively young and new technology is developing so quickly, both the content provider and the customers' requirements remain incredibly fluid. So it is very important to do extensive up-front evaluation prior to a large labor-intensive development effort, so that when finished, it doesn't represent a product no longer viable in the marketplace. In addition, often when we create a new feature for our service, it is because one of our content providers has specifically requested it. As I mentioned, the content providers have very dynamic needs, which adds to the complexity of the entire process. This situation has also led to equal weight being given to project management, design and development, and testing groups. These decisions are always made as a team.

This brings me to the subject of what skills are needed by today's engineering professionals. Regardless of which discipline they pursue, I think a real emphasis should be placed in a few areas in particular. The first is thinking from the customer's perspective; second, developing a broad awareness of the tools and products in your profession, and finally, understanding the importance of the product features versus time-to-market trade-off. These skills will allow a designer to make quality decisions, since they will be seeing the larger business perspective, and not just one small piece.

All Aspects of Design

As with any engineering design concept and process, the ultimate goal is to improve the development time, quality, and cost of the intended product, be it a house, an alarm clock, or a toy truck. To accomplish this goal, a product must satisfy many objectives including reliability, appearance (aesthetics), ease of assembly, ease of maintenance, safety, disposability, and many others. In the best design practice, you consider all of these aspects at the earliest stages and continuously review them throughout the design process.

The three elements of quality, development time, and cost are all necessary to ensure the success of a project, because these elements determine a company's competitiveness:

- Timeliness: Can the product be delivered to the market in a timely manner that it is demanded by customers?
- Quality: Will the product work as intended and provide reliability and value?
- Cost (affordability): Do customers have the ability and desire to buy the product?

In business and industry, this trio of elements is generally referred to as *faster, better, and cheaper.* These three dimensions largely determine the value of a product to the customer. In addition to those primary goals, we must also consider the larger issues and factors that will contribute to the best possible design. These larger issues/factors include the social/societal impacts of the design; environmental impacts of the design; and ethical issues and consequences as a result of the design.

These factors will present themselves regardless of whether or not they are planned and accounted for in the design process. But if they are not considered, there's often unintended negative consequences associated with the final product in these areas. Usually the positive consequences are always identified because they will help sell the product! One classic example of this is the automobile. Although it has provided tremendous benefits to economic growth and personal freedom and travel, there have been numerous unintended negative consequences. For instance, although automobiles have allowed for the growth of the suburbs, they have also contributed to the deterioration of many cities. The negative impacts the automobile has presented to the environment is also well-documented and is evidenced in oil and gas spills, and health problems from air pollution. The list of advantages and disadvantages goes on and on. But these are some of the important issues that must be weighed and balanced when designing any new product or planning a project.

Industry Scenario

Industry Research Indicates the Necessity and Value of a Whole Systems Approach

Many recent studies by various industry and market research firms have been asking the question, What are the most important challenges facing today's designer/engineer? The responses from the engineering community are very relevant to the discussion of whole systems and the engineering design process. Today's engineering professionals have more responsibilities than ever, including project management, project budgeting, testing and evaluation, customer support, working on cross-functional teams, and computerizing the design function. Additionally, the designer/engineer has to keep up with changes in technology, such as CAD, informational databases, standard office software applications (word processing, presentation, graphic, and spreadsheet programs), understanding and working with new materials, industry/market trends, and much more. Because of these factors, less and less engineering professionals report that they are focusing on just one aspect of a project. The overwhelming trend is toward involvement in the total project.

The engineering community is also reporting a rapid increase in the use of the Internet for obtaining industry data, supplier information, new product and material information, and e-mail capabilities among customers and suppliers. The conclusion from these surveys all seem to indicate that the knowledge, skills, and abilities required in today's engineering professional focus on the individual who can juggle multiple responsibilities, possess a broad skill set, can adapt to change, is willing to continue learning, and is able to perceive a project from a whole systems perspective.

Summary . . . Links to the Whole

As the complexity of technology advances, so will its interdependence, requiring individuals with skills in designing and managing projects from a holistic viewpoint. And finally, given this technological interdependence, design choices and decision making become more critical, once again emphasizing the need for a whole systems mentality. Related engineering concepts such as concurrent engineering, life cycle engineering, integrated product development (IPD), knowledge-based engineering (KBE), and total quality management (TQM) all address the need for integration of people and processes for development of the best possible product.

Business and education both emphasize and value the skills and knowledge associated with whole system thinking. The whole system philosophy is the basis for the most engineering design process models utilized in business today. The ability of an individual to view a situation from an integrated, holistic perspective is one of the most valuable assets a person can possess. It allows an individual to consider all aspects of design, all aspects of the business, and when working as a member of a team, arriving at optimal solutions to design problems, which are driven by the customer and the marketplace.

Therefore, whole systems thinking (1) understands that a system is a collection of parts that interact with each other to function as a whole; (2) realizes that a change or omission in any single part *will* affect the entire system; and (3) understands that there is no single "right" answer, but that there is a process that can be used to minimize unintended consequences to arrive at the optimal solution.

TOOLBOX

An Introduction to Whole Systems Thinking

At the end of each chapter will be a list (like the one that follows) that is designed to provide a summary of the "tools" that you acquired in the chapter. These can be added to your mental "toolbox" for use in subsequent chapters, or anytime you see a positive application. For Chapter 1, the tools consist of:

- Taking a holistic approach to problem solving
- Questioning everything (ask and understand the right questions)
- Defining all aspects of a problem before developing a solution
- Understanding dependencies and interrelationships
- Concurrent engineering model
- Life-cycle engineering model
- Integrated product development model
- Knowledge-based engineering
- Total quality management
- Causal loop diagrams
- The nine steps of the Engineering Design Process

A-Team Scenario

A-Team Puts Together a Team of Experts to Design a Pump

As often happens in areas where there are a multitude of high-tech companies, small start-up groups spin off and form new companies focused usually around the development of a product.

Such is the case of the group that calls themselves "The A-Team Design Group." Throughout the textbook, we will be following the successes and challenges of the A-Team Design Group as they work through a real-life project and encounter the very issues that are presented in each of the chapters. This new company consists of five people who have brought a wide variety of skills to a new venture. The main objective will be the design and development of new product(s) for individuals with patent ideas.

The A-Team Design Group will provide the following services: prototype design work, Rapid Product development and analysis, market research, cost analysis, presentation packages for marketing, and final engineering production plans.

The in-house services will be the prototype design work, presentation packages for marketing, and final engineering production plans. The other services for rapid product development and analysis, cost analysis, and market research will be subcontracted out. For the customer, all these services will be coordinated through the A-Team Design Group office.

Let's introduce the partners in the A-Team Design Group.

Joe

- A senior Design Engineer with twenty years of experience in a wide variety of companies. His main credentials are in mechanical and electrical design. Joe's strength is his innate mechanical ability, but he has limited computer knowledge.
- Long on experience/low on technology

Carlos

- A CAD designer with fifteen years of experience in two different firms. One company was a large corporation with many projects and Carlos had mainly one type of job as part of a design team. The other company he worked for was a small consulting engineering office. Carlos wore many hats there and was part of a cross-functional design team.
- Short on diverse experience/long on team projects

Claudia

- An industrial designer with ten years of experience in a corporate environment and seven years as a private consultant. Claudia has many connections through her seven years of consulting. She joined this design group because she felt the engineering strength brought to the group by Joe and Carlos would be a good balance to her ability to connect with clients.
- High on networking skills/low on engineering skills

Russell

- A technical illustrator with a university degree in computer science. Russell has worked in industry for ten years for three different firms. His background in computers will place him as the technical wizard in this design group.
- Short on experience/long on technology

Shantel

- A researcher and planner with fifteen years of working for a large defense corporation. Shantel prides herself on her organizational skills and ability to locate information. She hopes to expand her skills with the new group in the area of project management and market analysis.
- High in planning skills/low on experience of working in a small group

A-Team Scenario
continued

What brought this group together was an idea for a new kind of mechanical pump that a man named Earl had conceived. One day, Earl drove down from the north country with his truck. A patent lawyer in the city nearby had referred him to Joe. Earl knew he needed to do something more than just talk about his idea in order to get it out on the market. It was a good idea for a pump. This mechanical pump could be used in places where most other styles of pumps couldn't fit. It could endure extreme temperatures and could be modified to retrofit into existing mechanical setups.

Earl knew this pump would work. He had a local firm develop a working model. All his ranch neighbors came to see it and of course wanted one. So now Earl was on his way to see Joe with the only model of this pump in the back of his truck.

Joe met with Earl and realized that there was more to this project than just the development of a set of engineering drawings. At this point, Joe realized a design team was required to help Earl fulfill his dream of manufacturing his pump.

The A-Team Design Group was about to enter the picture. Earl met with the A-Team and demonstrated the one functioning pump and explained all the potential ways this pump could be utilized. The team took pictures, copious notes, and asked many questions. Earl, not ever having worked with a design group before or taken any product idea to market, was not sure of all that had to be done. He did not know how to answer some of the group's questions. The group realized there was a great deal of research that would need to be done on this project to get these questions answered. What Earl knew is that he needed a marketing plan, a set of engineering drawings, and a market feasibility study to qualify for any venture capital. But what Earl felt he really knew was what his pump could do. Earl reluctantly left his one pump and sketches in the hands of this newly formed design company and drove back to his ranch; his head filled with plenty of technical stuff.

The A-Team Design Group reviewed with Earl all the specifications that he wanted to see built into this pump.

Design Specifications

1. This pump is to pump irrigation water or pesticide liquids.
2. The pump should be small in size so as to easily be attached to a farm vehicle or truck.
3. This pump should be designed to be able to function in different orientations so it can be attached to a multitude of different vehicles and in different places on the vehicle.
4. The pump should be able to handle high volumes of fluid.
5. The pump should be able to withstand extreme heat and cold temperatures.
6. The design and feasibility study should be complete in six months.

Respond to the A-Team scenario by answering the following questions.
Use any tools or ideas you have learned in this chapter.

1. What are some of the considerations the A-Team needs to make prior to beginning this project?
   _____
   _____
   _____
   _____
   _____

Design Tools for Engineering Teams

A-Team Scenario
continued

2. What different elements of a whole system design are in place with the A-Team Design Group?

   _____
   _____
   _____
   _____
   _____
   _____

3. What tools or expertise do you see in the A-Team Design group that will support the development of this project?

   _____
   _____
   _____
   _____
   _____
   _____

4. What potential stumbling blocks do you detect that could stand in the way of getting Earl's pump to market?

   _____
   _____
   _____
   _____
   _____
   _____

5. What questions would you ask Earl about his pump?

   _____
   _____
   _____
   _____
   _____
   _____

6. List what needs to be done next by the A-Team Design Group.

   _____
   _____
   _____
   _____
   _____

Chapter 1 | An Introduction to Whole Systems Thinking

TERMINOLOGY & DEFINITIONS

causal loop diagram — a visual tool designed to allow quick analysis of a cause and effect situation.

concurrent engineering — an integrated approach to design, production, and customer service that emphasizes the advantages of simultaneous, or concurrent, product design by utilizing individuals from various areas of the business in the up-front concept and design phase, with special emphasis on customers and their needs.

Engineering Design Process — a sequential series of steps or actions that a designer takes when solving a problem or creating a new design.

holistic — related to or concerned with wholes or with complete systems rather than with the analysis of, treatment of, or dissection into parts.

integrated product development (IPD) — the process of developing a product such that all sub-processes required for that development are integrated seamlessly. By that, it is meant that a product should be designed and developed to satisfy all of the conditions a product will encounter, not just the initial ones.

knowledge-based engineering (KBE) — the concept of developing computer models to simulate the best-known engineering processes. Typically, KBE is used by creating a set of engineering data, comprised of components such as CAD/CAM/CAE data, manufacturing data, tooling data, and structural information. This data is then used in an integrated manner to develop a more detailed and comprehensive design plan, while also allowing for better analysis and evaluation throughout the process.

knowledge management — a system that applies an organization's knowledge to its decisions with employees working hard to represent it, transfer it, make it accessible, and encourage its use.

life cycle engineering — an engineering design model whose premise is that the entire life of a product should be evaluated at the beginning of the design process. The optimal design cannot be achieved unless performance, costs (both initial and ongoing), reliability, maintainability, disposability, and market trends are addressed up front.

system — a regularly interacting or interdependent group of items forming a unified whole; a group of interacting bodies under the influence of related forces.

TERMINOLOGY & DEFINITIONS

total quality management (TQM) — a philosophy that calls for the integration of all organizational activities to achieve the goal of serving customers. It seeks to achieve this goal by establishing process standards, maximizing production efficiency, implementing quality improvement processes, and employing integrated teams to more effectively design customer-driven products.

whole systems thinking — the ability to conceptualize problems in a manner that accounts for the entire structure/system, thereby leading to a fundamentally sound solution, or, realizing the interconnectivity of all things and the corresponding effect one action has on the system as a whole.

Activities and Projects

- **Describe** the potential desirable and undesirable social, cultural, political, and technological implications of the following scenario, using a whole systems approach.

  As an architectural designer in a small American city, your firm has just been awarded the commission to design the city's first multistory office building for a large client who is relocating its corporate headquarters. The client feels that putting its corporate offices in a small town will give its customers the image of a company that believes in more traditional American values and way of life. However, the community is apprehensive about a large company moving into town, with its associated problems, such as increased population, traffic, pollution, and increased needs for services, such as schools, police and fire protection, water, and utilities. What could you do, as the buildings designer, to help accomplish the client's goal and address the community's concerns? Are there any inherent contradictions that should be analyzed? What technologies and design aspects could you employ (i.e., building style, materials, location, etc.) to help satisfy both the client and the community? Make a list of the potential positive and negative outcomes for the community, the client, and the architectural firm.

- **Assemble** a design team to examine a potential project in your field of engineering study (architectural, mechanical, civil, or electrical). Then, taking a whole systems approach, develop a proposal that addresses as many aspects of the project as you can think of. Put the final solution in presentation form and deliver it to the class for peer review and comment.

- **Identify** at least two relationships that could be used to illustrate a causal loop diagram. Attempt to identify one relationship that shows positive feedback, and another relationship that shows negative feedback.

- **Select** a project related to your specific design discipline (architectural, civil, mechanical, or electrical) and apply each step of the engineering design process to the project. List the potential issues that would need to be evaluated, including aspects of both design and business.

- **Select** a common household product, evaluate and redesign the product for possible improvements. Evaluate and document the design trade-offs of timeliness, quality, and affordability. Along with listing the potential benefits, list the possible negative effects of the product to society or the environment.

- **Evaluate** the redesigned household product from the previous activity in terms of all aspects of the business. Specifically, determine the market for the product, production issues, financial considerations such as development costs, product pricing, and patent issues, and customer service/support plans.

Teams As a Tool in the Engineering Design Process

CHAPTER 2

> "Individual commitment to a group effort, that is what makes a team work, a company work, a society work, a civilization work."
>
> — Vince Lombardi

Introduction

The days of working on only one facet or part of a design project are behind us. Today the success of an organization is based on all members having the same knowledge, tools, and focus to make decisions for the good of the whole organization. This is called teamwork. *Design News*, an industry trade magazine that surveyed the engineering community, found that more than 80 percent of the respondents reported that their companies use cross-functional teams to develop new products.

This concept of team engineering has been driven by many factors. First, a global economy has brought different companies or divisions of a company to use a team approach to better communicate and problem solve. Second, downsizing of corporations has forced designers and engineers to do multiple tasks and have a broad knowledge of the design and manufacturing processes. Last, the development of technology has brought our society out of the Industrial Revolution into the Information Age. Access to team members through the Internet and the virtual speed of communication have directly changed how engineering can be performed. A drawing is now an e-mail away, and this access to technology allows for real-time engineering while bringing the team concept to a global market. The days of working in isolation and waiting for decisions to move forward are gone and this makes it easier for a team to produce a better product in less time.

> ABET Engineering Criteria 2000 includes the ability to function on multidisciplinary teams for engineering graduates to demonstrate.

Industry Scenario

Electronically Linked Teams Design the Defense Systems of the Future.

— *Rick DeMeis, Associate Editor,* Design News

Government defense budget cutbacks have forced the United States Armed Service to be "lean and mean," even in their basic approach to developing next-generation defense systems. One example is the development of the Army's Crusader cannon artillery system, a project in which engineers, purchasing professionals, and business managers work closely together to deliver "best-value" technology.

Key to efficiently getting the most bang for the defense buck has been "integrated industry and government product development teams" (IPTs).

"The system can provide real-time data exchange and configuration management by a variety of modes: T-1 line, the Internet, or dial-up," notes Flach, Crusader Business Manager for United Defense. "A variety of development tools are available to handle this information, depending on discipline."

Before the IPTs were tooled up and running, team structure and functioning were tested and proven. This took place in 1994 and 1995 at the Minneapolis site. The electronic "tool sets" were identified, selected, tested and proven for use at the remote sites. This program also formulated procedures and requirements for subcontractors to meet in bidding for program procurements.

Flach says integrated teams are much more effective than traditional government, contractor, and subcontractor relationships. "There is no alternative to a common development environment," Flach believes. "We are achieving milestones at a faster rate than most Department of Defense projects. Systems integration and interface control are realizing the most benefit, along with reducing the review time between organizations, since everything is in real time."

Integrated environments allow fast component design maturation.

This chapter will explore the advantages and disadvantages of team development and functions. Teams will be discussed as a valid method for design engineering. This chapter will provide you with information on how to participate as a design team member; understand the role of the team leader; determine how successful teams really work while valuing individual differences; and discuss the benefits of teams in a whole systems approach to design engineering. Upon completion of this chapter, you should be able to design an engineering project using teams as an integrated component of the engineering design process.

Orientation to the Concept of Teamwork

Team Definition

As we see it, a team is a collection of individuals, each with their own expertise, brought together to benefit a common goal. Hewlett Packard defines interactive teamwork as "individuals who effectively use leadership and group skills to maintain and promote cooperation and productive work relationships."

If we combine these two definitions, a team becomes a group with a common purpose who achieve a specific goal using the skills of each individual and mutual cooperation to produce the end product.

Utilization of Teams in Industry Today

Due to the changing scene of the American corporation and modern production, managers have favored the use of teams in almost every aspect of design, manufacturing and construction. The work styles that characterize how work has been done in our corporations find their roots in Adam Smith's prototype for productivity — the principle of division of labor. Smith's observation was that a number of specialized workers, each performing a single task in the manufacturing process, could increase the level of productivity more than if each individual performed all the tasks done. This principle of division of labor has been a model for production and engineering for decades. And it still seems to make sense, at least on paper. But given the fact that the shift in engineering is to process engineering, Adam Smith's principle that focuses on individual tasks without the sense of the whole process is not compatible to the changing scene of today's manufacturing and engineering realities (see **Figure 2.1**).

Industrial Age thinking: This is my job.

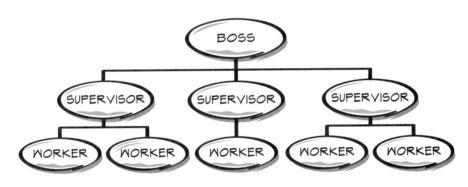

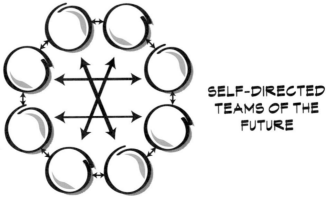

Figure 2.1
Historical view of working styles

Industry Scenario

Teams Making Sense for Two P.C.B. Designers

— *John Reynolds and Reina De la Fuente, Engineering P.C.B Designers, Fort Collins, Colorado*

One of the attributes people look for in potential new hires is the ability to be a team player. The ones that go into an interview boasting "I did this and did that and the company couldn't have done it without me," are the ones to be eliminated first these days. Why is this? Because nobody can get to where they are by themselves. The environment in which there is the most success is one in which there is support, open communication, and the ability to collaborate with all co-workers. We all are responsible for our assigned duties. Many times in order for teamwork to be truly effective, we must go beyond those duties to gain true satisfaction in our performance. Each individual must have motivation, self-awareness, empathy, social skills, and be capable of managing their emotions/egos.

Many companies much like our own offer team-building classes and workshops. These help to make people more aware of the dynamics of success in the workplace. We feel very fortunate in an environment where we are respected by and attuned to our co-workers and management. It takes effort on everyone's part to meet a deadline, develop a new idea, and troubleshoot a crisis. When everyone contributes, it gives a lot more meaning in a day-to-day job. We're always striving to improve our production (quality and quantity) and it always begins with individuals taking pride in what they have to offer and being able to communicate this effectively. Cross training is an excellent source of utilizing people's skills, while making them aware of how their work affects another. Our peers, as well as ourselves, hold us accountable for our work.

Teamwork can be broken down into two different types. One is specific to a project, where each member is responsible for a particular assignment within the project. The conclusion of the project is dependent on each individual. Indirect teamwork would involve a wide variety of sources for an ongoing objective. Resources from other areas would need to be brought in to accomplish the goals. Charting team progress and setbacks can be demonstrated in diagrams, data sheets, and flow charts. Round-robin discussions of everyone's input is critical to the development of the team.

Employee Empowerment

The shift to process engineering and team involvement creates a work environment that values and empowers the worker, breaks down the hierarchical structure of management, and provides connectivity to all aspects and elements of the design and production processes. The competitive demands on a company in today's market economy require maximum productivity from each employee. With the team format, employees have the opportunity to contribute to the success of the company, the project, and to themselves personally. Individuals have their input in the decision-making process and help provide solutions to problems, resulting in more personal responsibility and satisfaction.

As a student of engineering or a related technical area, the challenge is to build skills in communications, problem solving, decision making, and team participant skills.

This challenges the very core of America's standards for individualism, standards of personal achievement and definitions of success. Several years ago quality circles in Japan made news in the business world, but we've had the foundation for teamwork all along. We can look to sports and the military for successful models for teaming. It is now becoming commonplace in engineering, construction, and manufacturing.

Types of Teams and Their Functions

There are many types of teams, but the type of team used is ultimately formed and defined by the job it is to accomplish. Each team may have a different focus, yet they share the horizontal network that crosses different areas of a company. Today's company rates the ability to be a team player as the top workplace value.

Work teams: This is a team made up of individuals who work in the same unit and/or share the same function. They work together to produce a result or product. They have a shared goal.

Cross-functional teams: This team is composed of members from different departments or functions, and their job is to work together on projects that have combined job functions. The success of this type of team depends on the development of shared goals and understanding of each job function.

Task teams: This type of team is brought together for a specific amount of time to solve an organizational problem. They may comprise many different department representatives or just one unit.

Ad hoc teams: This type of team is less formal. They are usually formed to work on a specific task that is out of the realm of the regular team.

Development of Teams

The formation and structure of a team is based on the team objective, the operating methodology, and a commitment to a job or product. The team is composed of experts that are capable of adjusting to changing demands and needs as dictated by the job to be accomplished. Sound like a lot? Yes, but it's not impossible.

Traditionally, employees have been placed somewhere on an organizational ladder. They knew who their boss was, their job description, and the extent of their authority. As typical in an organizational chart, the boss was at the top with all the workers below. The chain of command and requests moved to the boss for decisions and down to the workers with directions (see **Figure 2.2**).

Design Tools for Engineering Teams

Figure 2.2
Traditional organizational structure

Today this same worker, designer, and engineer is empowered by sitting together on a team that represents many areas of the corporate organization. Each are presenting their own original ideas, and at times, taking on roles of leadership. In addition, the workers sit-side-by side their original boss who now is facilitating instead of directing (see **Figure 2.3**).

Figure 2.3
Team format

People in leadership support the team. For success, a person has to lead instead of boss.

Chapter 2 | Teams As a Tool in the Engineering Design Process

One of the first things individuals must do before participating in a team environment is to look critically at their own work values, communication with co-workers, concepts of achievement, and feelings about management. This change in format requires people to understand themselves, their strengths and weaknesses, and the talents they can bring to the team. It also requires the team to accept and acknowledge the strengths and weaknesses of each team member and ultimately trust and support each team member until the end result is accomplished. As a team develops, each person will have to ask himself or herself, Is my vision for success the same as the team's?

For a group of people to become a team they must have a shared vision. A vision is what the team wants to accomplish together. This vision is what will motivate the team to work together through good and bad times to meet their goal. A team coach can inspire the vision somewhat like the coach of a winning sports team. This vision can also come from the team members themselves. It is the vision that brings a team to high levels of performance and inspires commitment. An example of a team vision might be for the company team to outproduce all other teams in the company's history.

To create a team with vision takes hard work, and all team members must accept the challenge of the vision. The vision must be in keeping with the organizational goals of the company and the team must have a plan of action to obtain the vision.

Team Success

It is a two-way synergy between each team member and the team as a whole. The success of a team depends on this synergy. A team can be successful if the members open up and join into the process (see **Figure 2.4**).

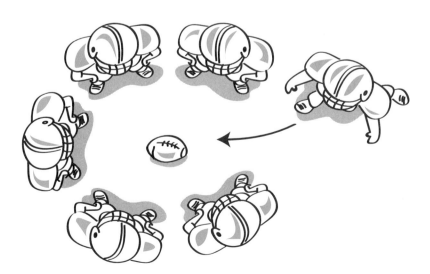

Figure 2.4
Successful teams utilize contributions of each team member

Each member must trust the others and support any of the team members in time of need. This requires that the team members recognize when help is needed as well as know when to ask for help.

Team success is also dependent on effective leadership whether this leadership is in the form of self-management or is facilitated by a designated group leader. Both have a place within the realm of design and engineering. Again, the structure and end result of the team organization is determined by the required outcome, the company comfort level to self-directed teams, and, in some way, the level of understanding of how to function as a team.

Education and training is critical for all members of the team. Companies are providing employees formal training in a variety of topics that support team development. Some important areas for a team or team members to be trained in could be team dynamics, conflict resolution techniques, or team leadership development.

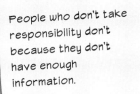
People who don't take responsibility don't because they don't have enough information.

Team failure usually occurs when management does not understand the concept of a self-directed team, or a team member is not a team player and has a personal agenda. If all the team members do not understand how a team works or accept the team goals, failure is sure to occur.

Team success also occurs when all team members understand the benefits of the team process. This realization helps individuals to let go of their personal agendas and join the team (see **Figure 2.5**).

Figure 2.5
When a team builds interdependency, it becomes stronger than any one individual. This is synergy.

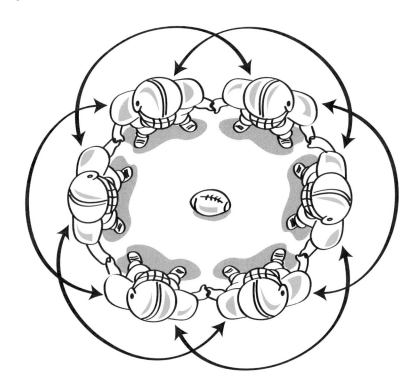

Once a group of individuals progresses and evolves to a team with a vision, each individual will feel the benefits of being a team. Listed here are just some benefits a team could experience.

Benefits for the individual and the team

1. Workloads are shared.
2. Cross training eases the workload as well as gives the individual additional employable skills for growth and advancement.
3. Chances for leadership and personal satisfaction come from being valued for their individual input.
4. Sense of belonging to a successful process.
5. Ability to accomplish more than if working independently.

Team Empowerment: How Is It Established and Maintained?

Empowerment is the desired end result for both the individual and the team. Empowerment encourages employees to participate actively in the decision-making process. It allows them to achieve recognition, involvement, and a sense of worth in their jobs, thus improving job satisfaction and morale. By working together, employees increase their power, develop cooperation, and build trust, so that the team can establish common goals and take risks together, with personal initiative and creativity.

A team is a group of individuals that bring together all different levels of expertise, sensitivity, and a willingness to participate in the team process. This is a challenge to most designated leaders. The charm of success will lie in the conversion of individual values and beliefs into team values. This is accomplished through a process called norming. The objective established in the norming process must be measurable and realistic for the team to succeed at their mission.

> Rule of Thumb: Be participative when you can and as often as you can.

Development of Group Norms

Norms can be described as a set of values that are based in beliefs and are acted on through specific types of behavior. Norms are known and established behavior. For example, being quiet in a hospital is a well-established norm. A team will follow very specific steps to establish group norms, but the first step begins with the purpose of the team. Value building and norms will follow.

Step 1: Identify the mission of the team

This is the very beginning stage when the team first gets together. The team or facilitator is assisting in identifying the following:

- What is the role of the team?
- What does the team have to do?
- How will they accomplish the task?
- Why are they doing it?
- What resources are available?

A mission statement clarifies the task to be done. Building motivation starts with a vision. The vision of the job directly affects performance. A team's success starts with a shared vision.

Step II: Establish team values

The next step is to identify team values. All team members should establish a list of what is important to them as individuals and therefore what they can contribute to the success of the team. Values are fundamental truths that would not likely change over time (see Figure 2.6).

Top of the Mountain

A team with a clear mission will accomplish the task. A group hike to the top of a mountain looks and feels very different from a rescue team whose mission is to locate stranded hikers on top of a mountain.

Figure 2.6
Example of team values

The list of values should be short and concise, and in alignment with the company values. Team members must come to a consensus of agreement that all values are for the benefit of the team, even if an individual may not believe that value is important to them.

Step III: Establish group norms

The group norms will be the guidelines for how each team member treats one another. These norms should reflect the values the team has adopted. Some call these core values, but basically they are rules that every team member agrees to follow because they are created by the team.

Norms are very important because they govern the behavior of each individual and ultimately the group. Norms often are already in existence in a group, for example, the most vocal person will automatically become the leader, or it is traditional for one person to show up late to a group meeting. If these unspoken types of norms are not acceptable, it is important that they be vocalized in the norm setting stage of team development or they will continue and erode the effectiveness of team communication.

Successful teams do not leave norms to chance. Successful teams have clear, documented norms that establish the behavior for each team member.

Suggestions on How to Establish Team Norms

- Create a list of norms by brainstorming. No norm is a "stupid" norm.
- Look at each norm and talk about its impact on the team and team goals.
- Identify the key norms that everyone can come to consensus on. Keep this list relatively short so it is workable, but if the norm is important, it should be included.
- Establish consequences if norms are broken. People want things to be fair.
- Take time in this process. Give each team member time to contribute their ideas and time to think over the identified norms before final adoption. Remember: All have to work by these norms, therefore, all must agree to agree.
- If a team member has a dissenting vote that is in contrast to the rest of the team, find out what would be necessary for this person to give total commitment to the process. Adjust the norms to work if necessary.

A successful team will only function when all team members cooperate with and commit to all other members. An outside facilitator may be helpful to bring the team through the norming stage.

Each team is unique unto itself and the norms it establishes will reflect this uniqueness. For a team to be successful there are some general norms to consider. For example, get input from all team members, even the quiet ones. Silence is a strong enemy that can undermine a team effort. It is also important to establish the method for team agreement, for example, consensus versus majority rule. This can be a norm.

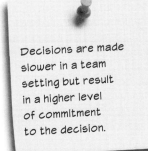

Decisions are made slower in a team setting but result in a higher level of commitment to the decision.

Some other good norms to consider are scheduling of team meetings, written agendas and minutes for each meeting, and how often the team should meet.

Norms can also address important concerns around confidentiality issues or how the team will handle conflict.

Norms are the framework for how a team will function. If the norms are carefully established and the team reaches consensus on them, the team has a better chance of succeeding at the task it has been given.

It is recommended that a copy of the norms be given to each team member and placed on file with the company.

Step IV: Identify strengths and weaknesses of each team member

The team is about to embark on the path toward the development of a product or idea. Roles and responsibilities are being assigned and distributed. It is at this time that the team needs to identify their individual strengths and weaknesses. This is what makes the teams work together. A team can corporately work together to balance the weaknesses of the group and to identify and utilize the strengths. It is like being on a survival camping trip and knowing what expertise the group can count on and what are the limitations. If only one person on the camping trip can read a map to get the group back home, then that person is given the map-reading responsibility.

Each team member should, at this time, list his or her talents, skills, and limitations. This process helps identify the need for cross-training, job duties, training needs, and most importantly, an understanding by the team of their corporate strengths and weaknesses.

EXAMPLE OF A TEAM'S STRENGTHS AND WEAKNESSES

Team	Strengths	Weaknesses
Team Member A	• Experience with various aspects of cars • CAD: solid modeling	• Design analysis (i.e., FEA) • Knowledge of tolerances • Poor handwriting
Team Member B	• Off-road racing and mechanic experience • Creativity	• No FEA experience • Report writing/formatting
Team Member C	• Manufacturing skills and experience • Strong interest in the project	• Time management • Too persuasive
Team Member D	• Analytical and solid modeling • Math skills • Work Experience	• Delegating work • Physical interpretation of analysis and calculations
Team Member E	• Organizational and leadership • Attention to detail	• Lack of ability to compromise • Computer knowledge • Short attention span

The other outgrowth of this exercise is to discover the differences that exist within the group. Some differences will be obvious, such as work skills; other differences may be harder to identify, such as work styles or organizational standards. If a harmonious team is to emerge from this process, each member must be truthful about their style and personality quirks. Once these are identified, the team must work at putting into place a plan that builds positively on the differences while still valuing each individual. This process was once the responsibility of the supervisor. The boss would reward individuals for their strengths or penalize individuals if their weaknesses interfered with their job. Now with the collaboration of the team effort, every team member's strengths become a support system for every team member's weaknesses. Again, the synergy of "the whole is stronger than that of its parts" comes into play.

Some key areas where differences will arise and may have impact on how the team can perform will be in communication styles, the way a person organizes their work habits, the level of expertise and training an individual brings to a group, and how individuals respect and respond to each other.

A few tips to help the team convert differences into strengths are:

- First identify the differences.
- Discuss how the differences can work against you. Then discuss how the team can use the differences as a strength.

    Example: Joe is shy, but very detail oriented. Consuella is overt and loves to talk and brainstorm. To ask Joe to represent the team in Corporate Think Tank meetings would be adverse to his personality, but Consuella would thrive in this environment. The two together would be a great team. Joe could document the ideas that Consuella comes up with.

- Knowing the differences helps you head off potential collisions. Discuss the team routines and try to foresee potential problems and suggest solutions before any arise.

The Cycles of Team Maturity

Given all the positive team-building work that has gone into developing a successful team, the reality is that a team will have growing pains and will develop at different rates depending on the team's experience in functioning as a team.

This whole process is called "norming and storming." Yes, the team has established norms, but until tested, the team will not know how well it will perform. Morale is generally high at the beginning. All the goal setting and norm building begins to bring a diverse group of people together. In reality, it takes more than discussion for individuals to change individualized approaches to work. So the second phase of team development can produce conflict or a drop in morale, and is a critical point in the team life cycle. This phase is known as "storming." During this

Design Tools for Engineering Teams

time, resentment surfaces, personality issues emerge, and often this is followed by disagreement. Conflict resolution is the tool to bring the team out of a "storming" phase. Training may be necessary in conflict resolution techniques.

There are many opportunities for training in conflict resolution, through your company or from private trainers. What is important here is when a team trains together it is an opportunity for all team members to hear the same information. The whole team has the same tools and aids to draw from to work through a storming session.

A team can emerge stronger from the storming phase or totally fail. Success or failure depends on the member's willingness to revisit the original goals and norms established by the team. Discussion on what's not working and positive ways to correct the situation will bring the team out of this phase, however, teams can remain in this phase for an extended period of time. This is the most difficult time in a team's life and it will take a strong vision and commitment to work through the storming phase.

It is important for the team to understand from the beginning that the storming phase will come. This knowledge helps the team recognize when it is occurring and to meet the issues head on. Storms can come more than once.

A team may reach a mature level of performance, but regress to an earlier level if a change has occurred such as accommodating a new team member, completing a task, or changing the team leader or coach. The team may need to re-norm and define its value system. Looking at the team norms is a good tool to pull out of the storming phase.

Team transition can lead to team success through focused assessment. Suggested questions the team could ask themselves:

- How is team performance measured and how often?
- What are the tools of measurement (are they the team's or the company's)?
- What rewards are in place for the team?
- How will individuals be evaluated?
- Who conducts the evaluations — team members, team leaders, customers, or company management?

When a team is scheduled to disband because of failure or if the job is finished or people are transferred out, the level of productivity and morale may be affected either positively or negatively depending on the circumstances.

A team becomes fully productive usually after a storming phase or two. Team changes can be smooth if the team looks for the positive and shares what they have learned from their experiences as well as celebrate their victories.

> We are programmed that conflict and differences are bad. A positive response to conflict is confrontation and resolution. A negative response is avoidance and attacking.

Chapter 2 | Teams As a Tool in the Engineering Design Process

Team Leadership and Team Control

Team leadership and team control can be a sticky discussion. It is real easy in the beginning to turn leadership over to those who already have the power or who are naturally the most vocal or assertive. But as the team matures, the control shifts from the leader to each team member. This is not a smooth process but gradually everyone will assume responsibility for its performance.

Figure 2.7
Transition of leadership in teams

45

The structure for team leadership will depend on the type of team. As shown earlier in Figure 2.1, the work group has a designated leader with working members. A self-directed team may rotate the team leadership. Effective teams will accept and work with the team design to be effective in attaining their goals. What is important for each team member to understand is the reporting line for the team, who represents the team to customers or management, or who is responsible for coordinating different functions of a team as in a cross-function team.

There are two different types of team leaders: a *team leader* who is appointed by the team, or a *coach*. A coach is usually a former manager. As a team initially forms, this is often the type of team leader that is in place. Gradually the coach will assume less and less responsibilities as the team becomes stronger and more secure, and the concept of a rotating team leader will emerge.

ROLE OF TEAM LEADERS

COACH

Lead the team from dependence on one leader to interdependence on the whole team

Manage conflict

Build synergy and team trust

Promote risk taking and help team members identify their personal strengths

Clear the path for the team to function

Listen and mediate

TEAM LEADER

Maintain a working role in the team

Manage conflict

Maintain synergy and trust

Monitor the team purpose, goals, and objectives

Facilitate the team meetings

Encourage risk taking and individual initiatives for the benefit of the team

> When power is given to individuals to run a business, it frees up management's time to build the business.

Chapter 2 | Teams As a Tool in the Engineering Design Process

Industry Scenario

Teamwork Gives Maytag a Jump on the Competition.

Maytag's Galesburg, Illinois, facility may be 88 years old, but it is using some of the most innovative product development strategies in the industry. In recent years, this Refrigerated Products division began joining engineering, purchasing, and suppliers early in the design stage. The results have been impressive.

"Supplier-integration benefits have resulted in about a 50 percent lead-time benefit with respect to time-to-market over the past three years," says Scott Giles, director of procurement at the Galesburg site.

Early supplier involvement also has helped Maytag integrate advanced materials into new products and develop new production methods, such as a more efficient precoating process for refrigerator cabinets.

The Refrigerated Products group used this innovative design strategy (and the subsequent manufacturing process) to design and develop a new line of energy-efficient refrigerators. The Advance Performance Design (APD) line, which includes a top-mount refrigerator as well as a side-by-side model, not only meets today's tough energy-conservation requirements, but also uses new materials and manufacturing methods that have improved quality, increased productivity, and reduced costs at the Galesburg site.

For example, with the supply team leader process in place, Maytag has been better positioned to tap a supplier's technical expertise, according to Giles. "What this means is that everyone now deals with his or her peers in other departments. No longer do we have a 'throw-it-over-the-wall' operation, but a gradual movement into the world of concurrent engineering. It involves suppliers at the outset of the design process so that we can make more use of their talents. In short, the roles of all the players are changing, based on their expertise." Moreover, added resources within Maytag assure that quality improvement escalates in conjunction with any design improvement.

Such teamwork also offers big benefits to engineers. No longer does engineering have to perform time-consuming testing and specification conformance approvals as its confidence in the certified suppliers' capabilities grows. This frees an engineer's time to concentrate on getting products to market faster.

Summary . . . Links to the Whole

Teams and teamwork can work positively to create synergy. When teams are successful they become an entity that is greater than the sum of its parts.

Therefore, when we put teams in the context of whole systems engineering, each small team becomes part of the larger system to produce a totally integrated process for engineering. This phenomena of teaming is occurring in companies large and small, within the four walls of the factory, to cyberspace teams. It is a major power shift in the engineering workplace that has changed how products develop and come to the marketplace.

Design Tools for Engineering Teams

TOOLBOX

What Makes a Successful Team

- Three factors critical to high performance teams
  - Commitment to a shared goal
  - Trust
  - Open communication
- Steps to develop a team vision
  - Each individual team member must self-examine their vision for the team to determine if they can commit to the team vision
  - Align the team vision with the organizational goals
  - Develop a plan of action to attain the team vision
  - Initiate the plan
  - Celebrate obtaining the vision
- Steps to develop team norms
  - Identify the team mission
  - Establish team values
  - Establish team norms
  - Identify strengths and weaknesses of each team member
- Ways to facilitate a team change
  - Look for the positive
  - Share what the team has learned
  - Review team norms
  - Plan for the future
  - Celebrate transitions
- Tools to aid the team coach
  - Be helpful
  - Demonstrate flexibility
  - Be open-minded
  - Listen actively
  - Assume a proactive mindset
  - Be the keeper of the vision
  - Keep a sense of humor
  - Be optimistic
  - Provide positive feedback
  - Keep quiet to hear the whole team

A-Team Scenario

A Design Team is Formed

As Earl's truck pulled away, Joe, Shantel, Carlos, Claudia, and Russell stood and waved goodbye. As he went out of sight, the five of them turned toward each other and stared at the pump and the few scant notes and sketches. What went through each person's mind at that point is unknown. What was verbalized got them started. Carlos blurted out, "We've got ourselves a for-real project!" Russell joined in, "Should be a snap!" At which point Shantel sighed, "Wow! We have a lot to get together and find. I'm not so sure this is going to be 'a snap.'" "OK, you guys. Let's not panic yet. We have a lot of resources out there to help us," piped up Claudia. Joe was quiet. As usual, he was thinking and studying the pump. Already he was making notes and sketches. As everyone was chattering away, Shantel asked Joe what he was thinking. "Well," Joe mused, "I think we do have a lot to plan and I believe we do have plenty of resources. Most of all we have our own resources." Everyone chimed in at the same time, "We do?" "Yep! Each of us has a skill, talent, or experience to pull this off. That's why I've asked you all to join me on the project. We are a team now. Remember we name ourselves the 'A-Team.'"

"Joe," questioned Carlos, "can I ask the group a question?" "Sure, Carlos, but you don't need my permission. I'm not the leader here. We haven't organized ourselves that way. So shoot, what's your question?" "Well, I have a lot of team experience and training from my former jobs. Has anyone else been trained in how to work in a team?"

A-Team Scenario
continued

Most of the group had limited teaming experience. They realized after discussing their project that they had better understand team dynamics. Carlos was helpful in identifying various aspects of the team process, but he wasn't comfortable training his peers. So the group decided to put themselves through some team training from a professional organization that specialized in corporate training.

As the newly formed A-Team began their process of gathering materials for the pump project, a team trainer was brought in to train the group in the dynamics of teaming. Following the training, the group scheduled a planning meeting. They learned from the trainer that there are specific steps a group must progress through in order to be an effective team. This is real important for this A-Team since not all of the members know each other. Joe had brought them together for their experience.

"Hi, guys," Claudia said cheerily, as she joined the group for the planning session. "Are you all ready to start norming, storming, and forming?" "Sure, but I don't like the sound of the storming part. It seems in every project I've been on, someone doesn't hold up their end of the bargain," retorted Russell. "Well, I guess if we follow the steps like our trainer said we won't have to storm," replied Shantel. "Hey, where is Joe? It is 30 minutes past the time we said we'd start. I know Joe knows more than all of us, but we sure need him to be with us." "Maybe he feels he doesn't need to do this process part on goals, norms, and our personal strengths and weaknesses," Claudia worried. "Should I give him a call?"

The meeting finally began with everyone there. The team developed goals and some group norms. Shantel helped set up a chart that listed each team member's strengths and weaknesses. This part of the exercise was really helpful, as well as interesting. The group was already beginning to form opinions about who should do what job for the project based on the surface information they knew about each other. The strengths and weaknesses chart allowed for each member to be honest about what they could contribute and where they were the weakest. Even Joe, who is the quietest among them, came forward with his thoughts. It was important to the group that everyone had a voice.

Carlos piped up, "Wow! I never would have guessed I was the only one with solid modeling experience in CAD. Guess that will be my main job." "That's okay with me," said Claudia, "I feel I can better benefit the project by finding and working with our outsource people. I know a great outfit that solely specializes in 'rapid product development'." Russell joined in the conversation, "You know, I really like our norm that says we will make all our decisions together." "I do too," Carlos replied. "Our norms seem like our code of ethics; something to help us stay on track." Claudia agreed, "Yeah, and their ours, not some upper manager person's code."

The group settled in to work on their respective jobs feeling rather proud of the work they had done together on forming their team.

A-Team Goals

- To design engineering plans for a pump design that will work in a wide range of temperature settings and can be retrofitted to different size farming vehicles.
- To conduct a market feasibility study to develop market potential for ranching and farming applications.
- To package all pump documents (engineering plans, cost analysis, mechanical test reports, market analysis report) into a presentation document.

A-Team Norms

- All major project decisions and outsourcing decisions will be made with all team members' input and consensus.
- All meetings will start on time.
- All team members will have their turn to voice concerns, report information, or suggest ideas.
- All ideas will be considered — no idea is a dumb idea.

Design Tools for Engineering Teams

A-Team Scenario
continued

A-TEAM'S PERSONAL STRENGTHS AND WEAKNESSES CHART

	Strengths	Weaknesses
Joe	• Experience in mechanical applications • Applied analytical math skills • Knowledge of pumps	• Too independent • No solid modeling CAD skills • Lack presenting skills • Team experience limited
Carlos	• Teamwork experience • Solid modeling • Multiple jobs • Creative	• Weak on test and applied problem-solving • Impulsive
Claudia	• Connected in the engineering community • Good presentation skills • Enjoys working in a group • Can see the big picture	• Weak applied engineering skills • Doesn't deal with negativity real well
Russell	• Good presentation skills • Computer skills • Rapid product development student	• Not very creative • Time management • Report writing
Shantel	• Organizational and leadership skills • Report writing • Loves to research	• Impatient • Hates tardiness • Doesn't like change

Respond to the A-Team scenario by answering the following questions. Use any tools or ideas you have learned in this chapter.

1. Based on the design specifications provided in Chapter 1, did the A-Team write goals that were inclusive of these specifications? If not, what would you add?

_____
_____
_____
_____
_____
_____
_____
_____
_____

Chapter 2 | Teams As a Tool in the Engineering Design Process

A-Team Scenario
continued

2. Did any team norms occur to you that would help the A-Team stay on course? (Review the definition of a team norm in this chapter.)

   _____
   _____
   _____
   _____
   _____
   _____
   _____

3. Review the A-Team strength and weakness chart. Are there any ideas that stand out to you that could present more problems for the project? After studying the strengths and weaknesses chart, what are some of the strengths of this team? Any other observations about the A-Team?

   _____
   _____
   _____
   _____
   _____
   _____
   _____
   _____
   _____

TERMINOLOGY & DEFINITIONS

norms — principles of right action, binding upon the members of a group and serving to guide, control, or regulate proper and acceptable behavior

process engineering — design, development, implementation, and support of manufacturing processes

self-directed teams — a group of individuals brought together to achieve a common goal, with shared values, purpose, mission, and accountability through participative leadership

synergy — when the unit or team becomes stronger than the sum of the individual members

values — fundamental truths and beliefs that the organization or individual uses as a guideline while accomplishing the vision

Activities and Projects

○ Complete a "Self-Assessment for Team Member Readiness" survey.

The following self-assessment will help you determine your areas of strength that you can contribute to the team effort and areas you need to improve on to become a fully participative team member. Answer as honestly as you can.

TASK BEHAVIORS	CIRCLE ONE
Do I delegate work to others?	Y N
Do I pass on information to others when it relates to a project or problem?	Y N
Do I reflect or review the results of a job before moving on to the next project?	Y N
Am I comfortable asserting my viewpoint in a group?	Y N
Do I make decisions without consulting others for input or advice?	Y N

RELATIONSHIP BEHAVIORS	
Do I share all my strengths or weaknesses openly with others?	Y N
When others are talking to me, do I listen first before giving advice?	Y N
Do I readily give a "pat on the back" for a colleague's good work?	Y N
Do I give others a chance to express their ideas and opinions?	Y N
Do I prefer to work alone?	Y N
Am I accepting of a differing point of view?	Y N

TEAM COMMUNICATION	
Do I listen to some people better than others?	Y N
Do I bury problems hoping they will go away?	Y N
Do I think discussion is an important communication tool?	Y N
Am I able to ask for help?	Y N
TOTAL ALL YES'S AND NO'S	___ ___

Activities and Projects (continued)

- Each question with a "no" answer indicates a need for further team training or personal self-adjustment. "Yes" answers indicate a strength or contribution you can make to the team effort.

ACTION PLAN

Areas of Improvement	Action to Be Taken
_____	_____
_____	_____
_____	_____
_____	_____
_____	_____
_____	_____
_____	_____

Areas of Strength	Contribution to the Team
_____	_____
_____	_____
_____	_____
_____	_____
_____	_____
_____	_____
_____	_____

Activities and Projects (continued)

- Write a mission statement for a team project in which you are involved.

- Reflect upon a group you have participated in or within which you are currently participating. List three to four norms that guide the group. Then reflect on whether these norms have guided or hindered the group to achieve its mission. Also indicate if these norms were established by the group members, determined by an authority figure, or if they were unstated norms that just developed.

- Research a local company that is utilizing the team format for their design or manufacturing process. Create ten to fifteen questions you would like to ask that relates to the team process. Base your questions on this chapter and on what further information you would like to know about teaming.

- Develop an assessment tool for evaluating the success of team activities. The following are examples of some types of questions to include.

    1. The mission of the team was clearly defined.
    2. Team members trust each other.
    3. Team members work together to solve problems.

Creativity and Innovation in Design

CHAPTER 3

> "*The scientist seeks to understand what is; the engineer seeks to create what never was.*"
>
> Theodore Von Karman

Introduction

Engineering is, at its very foundation, a creative activity. Albert Einstein once stated that, "The formulation of a problem is often more essential than its solution, which may be merely a matter of mathematical or experimental skill. To raise new questions, new possibilities, to regard old questions from a new angle, requires creative imagination and marks real advances." Today's competitive business environment demands solutions that are faster, cheaper, *and* better, and the only way to do something better with less resources, such as money, is to do it differently. This is where creativity and innovation enter the picture. But first of all we must define what is meant by creativity and innovation. By definition, creativity is the ability to produce through imaginative skill, to make or bring into existence something new, to form new associations and see patterns and relationships between diverse information. But more than that, and best of all, creativity is one of those elemental abilities that can enrich your personal and professional life. The creative process is fundamental to a challenging and rewarding life. Creativity's partner, innovation, is the term used for the process of transforming a creative idea into a tangible product, process, or service. Innovation is about improving the quality of a specific "thing," and allowing for more and better choices in our lives.

Having defined the importance and necessity of creativity and innovation, we must also remember that our culture is one of paradox. Although we hold up individuality, creativity, and innovation as some of the fundamental traits that make our society great, more often than not we educate and encourage children and adults that it is important to go along, compromise, fit in, and defer to authority. This very conformity destroys the habits that are necessary for creative thought to take place. Periodically, it is very important to stop and take the time to look at your world from a different perspective. The act of creativity is sometimes as simple as questioning your own assumptions, your own "truths" as defined by your life experiences, education, and culture. It is important to look at "obvious" answers and then use critical thinking and reasoning skills to determine if the answer really is "obvious." The renowned economist John Maynard Keynes once stated, "The greatest difficulty in the world is not for people to accept new ideas, but to make them forget about old ideas."

The most creative people always expose themselves to a wide range of perspectives — cultural, political, organizational, and personal. They understand that breakthrough creativity occurs when previously unconnected thoughts and concepts meet to form entirely new possibilities. Creative individuals tend to have a broad range of interests such as art, literature, music, business, and technology, which allows for more of the connections that promote creative thought. Imagination and creativity are the very heart of human experience. Even such subjects as engineering, logic, science, and mathematics are derived from original creative insights, and are fundamentally dependent on imagination and creativity.

> Chance discoveries do not occur by chance.

In today's global economy, information and technology is available to anyone, anywhere in the world. Because of this, creativity and innovation take on a whole new meaning. As John G. Young, former chief executive officer of Hewlett-Packard has stated, "Creativity is the only American competitive advantage left."

Creativity is a skill that can be taught and learned. Think of creativity as a habit. On a daily basis, you look for ways to realign things or ideas into relationships that did not previously exist. In the world of engineering, we can apply this same concept to the process called creativity, and we can then begin to find new and original ways to approach the world in terms of ideas, products, processes, and services.

Using creativity techniques can lead to more effective problem-solving skills that will facilitate a more productive and satisfying career in any profession. But problem solving and creativity is not easy for most people. It is typically a journey into the unknown. Problem solving begins with the acknowledgment of a need for change, or a sense that something is not quite right, and moves through a series of thoughts, feelings, beliefs, actions, and ultimately solutions. Creative problem solving is both an art form and a scientific process. It requires intuition and imagination. And as strange as it may seem, creative problem solving also requires careful analysis and step-by-step planning. This chapter is designed to provide a fundamental knowledge of both the techniques and principles of an integrated approach to creativity, problem solving, and innovation.

Chapter 3 | Creativity and Innovation in Design

Industry Scenario

— Bucky Wall, Program Management Training Manager, Microsoft, Inc.

Microsoft is known as one of the most successful companies in the world when it comes to envisioning and creating the future of computing software. As a Training Manager for the Microsoft Corporation, Bucky Wall provides the training, consulting, and networking necessary for the company's Program Managers to successfully deliver their standard-setting products to the customer. Making sure that Microsoft's Program Managers develop and maintain the skills required to operate in this most competitive of all industries requires not only an awareness of the latest technologies and a broad understanding of the industry, but the creative abilities to coordinate and deliver this knowledge to the right people at the right time.

In my position as a Training Manager for Microsoft, my first priority is to provide whatever is required for the company's various Program Managers to do their jobs effectively. Program Managers are the people responsible for delivering the finished product to the market. To understand how and why I perform my job the way I do, it is first important to understand Microsoft's philosophy regarding its employees. Very simply, we hire the smartest people we can find, and then expect them to adapt to a constantly changing world, a world that is changing technologically and culturally, both at the company and in the market. Also, due to these realities, most of our people will not remain in the same job for longer than twelve to eighteen months. It then becomes crucial to hire individuals who not only have a passion for their work, but who feel comfortable working on a variety of projects, who can view a problem from a larger perspective than themselves, and who are able to develop creative and innovative solutions to problems.

We also emphasize to our people that most of the problems the company will need to address to remain successful are business problems, not technology problems. One of my responsibilities is that of networking. By that I mean that one of my primary duties is to put the right people in touch with each other. Since the most effective problem solving, decision making, and innovation is not going to occur in isolation, it is vital that we put the necessary people together at the time it's needed.

Microsoft is fortunate in that we have more good ideas than ways to implement them. But issues such as time-to-market, distribution, pricing, shelf life, and other business related concerns are where creativity, problem solving and analysis become important. To address this reality, we operate almost exclusively in project teams. Individuals in these project teams must possess a cross-functional understanding of the entire groups roles and responsibilities. That way we can promote creativity within a specific plan of action. In other words, creativity and innovation can take place within a framework of the project's ultimate goal. Therefore it is very important that our people have a broad-based knowledge that allows for a comprehensive thinking process. Many times in the interviewing process for a potential new employee, we deliberately present abstract scenarios and questions for the candidate to answer to get a feel for their thinking process. Being able to successfully address these scenarios and questions, which are totally unrelated to the software industry, can provide us with a better understanding of how that individual will handle the unknowns and vagaries that come with developing products in a constantly changing technology and marketplace. So ultimately, in acting as a resource to provide people with the information they need, when they need it, my goal is to allow people to gain the knowledge necessary to turn problems into innovative solutions.

Qualities of the Creative Person

Every person on the planet has the potential to be creative. Each of us is a unique individual capable of creating. We are by our very nature, a creative animal. Even though all people possess the ability, those who are looked upon as creative have an awareness and a world view that others do not. Creative people also seem to have a unique ability to make connections and combinations of seemingly unrelated subjects to produce original ideas.

In his book, *How to Think Like Leonardo da Vinci: Seven Steps to Genius Every Day*, author Michael Gelb says that we all can unlock the creative genius inside us. The author states that there are seven critical principles that da Vinci followed and, regardless of whether you're learning a new language, studying to be an engineer, or just hoping to be more productive in your everyday life, exploring these principles will enable you to be more creative. Gelb has used da Vinci's native Italian to define the principles that allowed for perhaps the most creative mind in history to flourish:

- **Curiosita:** An insatiably curious approach to life
- **Dimonstratzione:** A commitment to test knowledge through experience
- **Sensazione:** The continual refinement of the senses, especially sight, as the means to clarify experience
- **Sfumato:** A willingness to embrace ambiguity, paradox, and uncertainty
- **Arte/Scienza:** The development of the balance between science and art, logic and imagination ("whole-brain thinking")
- **Corporalita:** The cultivation of ambidexterity, fitness, and poise
- **Connessione:** A recognition and appreciation for the connectedness of all things and phenomena, that is, systems thinking

Principles of Creativity

Using the seven principles as a foundation, here are some specific qualities that, if cultivated, can markedly increase your creativity:

- **The ability to see relationships and patterns.** This happens when you have command of more information to begin with. There are many aspects to creativity, but one definition would include the ability to take existing objects and combine them in different ways for new purposes. For example, Gutenberg took the wine press and the die/punch and produced a printing press. From art, music, and literature, to the daily chores of life, this ability is part of the nature of being creative.

Chapter 3 | Creativity and Innovation in Design

Industry Scenario

For a moment, close your eyes and mentally transform yourself back in time. It is 1941, in the forests of the Jura Mountains in France. A scientist is out hunting with his dog. At the end of a long day hiking the hills and valleys, he returns home, and, upon removing his jacket and pants discovers they are covered with wood-burrs. His dog's coat is also smothered with them. The scientist decides to examine them more closely using his microscope. What he finds is that the wood-burrs consist of hundreds of little hooks engaging the loops in the clothing and fur. That scientist, George de Mastral, decides to replicate nature by making a machine that will duplicate the hooks and loop model using nylon instead of wood-burrs. He called his new product Velcro from the French words for VELours and CROchet.

The rest is, as they say, history. Because a man decides to be curious, ask himself a question, do a little research, and make a connection, the world today benefits by its use of Velcro for thousands of different products and applications.

- A belief that you are **creative!** Do not underestimate the power of suggestion.
- The ability to look at a problem from **a different perspective or viewpoint.** Change your position when necessary, such as looking at a situation from a child's point of view.
- **Playfulness and humor,** be a dreamer. In developing his theory of relativity, Einstein had many dreams about the different manner in which time could operate.
- A work environment that is **flexible, open, and autonomous.** In an ideal workplace, an unusual idea or failure is not criticized.
- **Curiosity and a questioning attitude,** especially of fundamental concepts.

A different perspective is worth 50 IQ points.

What is a car? What is a house? What is a computer? The answer to an apparently obvious question can provide entirely new perspectives. For example, to ask the question of a designer . . . What is a house? The designer's answer to this fundamental question will provide the basis for the design. Is a house simply a place of shelter, large enough to accommodate a specific number of people, or is a house a structure that reflects an individual's environment, personality, and beliefs? Is a house a place that meshes with the natural world around it, or makes a unique statement within the landscape? Does a house indicate the owners' past or their future? As you can see, personal definitions of what a house is for can provide any number of possibilities. So often, a problem to be solved can be refocused or redefined to provide a new perspective just by asking basic questions. Albert Einstein was quoted as saying his key to success was simply that he "never stopped asking questions." Something else to consider is that creative individuals are overwhelmingly characterized by those people who ask more "why" questions, as opposed to "what" questions.

- **Imagery and Visualization.** Albert Einstein daydreamed about traveling on a sunbeam to the end of the universe to answer the question: "What would happen if one could follow a beam of light at the speed of light?" His imagery and visualization eventually changed everything we knew about time and space.
- **Subconscious thoughts.** Let an idea simmer while your brain makes connections. Don't move too quickly into a problem-solving mode just for the sake of having an answer. Even though quick solutions provide a sense of security, they also shut down our motivation to find the best answer. Creative people have the ability to remain "open" and reserve judgment until after sitting with a problem for a while. Subconscious thought acts as an incubation period for your problems and ideas that allow for making connections and links that will eventually provide you with many possible solutions.

Industry Scenario

The Value of Estimating . . . or, How Many Piano Tuners Are There in Chicago?

Many times in the world of engineering it is necessary to make a few quick "back of the envelope" calculations to determine the feasibility of a "creative" idea or solution. This ability to estimate is invaluable to anyone, but especially to the engineering professional. There is terrific advantage in the ability to arrive at quick and rough quantitative answers to unexpected questions about many aspects of the natural and engineered world. The skill of estimating was frequently used by one of the most widely creative physicists of contemporary times, Enrico Fermi. Fermi delighted in thinking up wide-ranging, arcane questions for his students and colleagues to discuss and to develop answers for. This ability required a deep understanding of the world, the knack of integrating everyday experiences, the skill of making rough approximations, inspired guesses, and statistical estimates from very little data. In other words, estimating requires a broad general knowledge of our world.

In a "Fermi question," the goal is to get a close approximation of the answer through reasonable assumptions about the situation, and not necessarily relying upon absolute knowledge for an "exact" answer. Here are some examples of what these typical questions looked like:

- How many drops of water are in the Atlantic Ocean?
- How many ping-pong balls would it take to fill the Washington Monument?
- How much human blood is there in the world?
- How many books are in an average bookstore?
- How big is a 1:1000000 scale map of the United States in inches?
- How long would it take to count to a million?

To look at an example of how the answer to a typical question could be estimated, let's examine one of Fermi's classic questions: How many piano tuners are there in Chicago? Well, let's look at the process and assumptions Fermi made in deciding the answer to this question.

To begin with, let's estimate that there are approximately four million people that live in Chicago. If there is an average of four people in a family, then there are approximately a million households. Let's guess that about one in twenty households owns a piano, so that's a total of 50,000 pianos in the city of Chicago (1,000,000/20 = 50,000).

Industry Scenario, continued

Now let us assume that a typical piano needs to be tuned about once every two years. That means about 25,000 pianos a year need to be tuned in Chicago. Let us further estimate that a piano tuner can tune about four pianos a day, which, in a typical work year (50 weeks x 5 days/week = 250), would permit them to tune about 1,000 pianos (25,000/250). Based upon our estimates then, there should be about 250 piano tuners (25,000 pianos /1,000 pianos per tuner per year) in the city of Chicago.

As it turns out, this number was within 10 percent of the actual number shown in the Chicago telephone directory. What makes this method work is based upon two important items. The first is mathematics; estimates that are high will, on average, be canceled out by estimates that are low. Second, and more importantly, a good broad-based general knowledge of the world is required, which has already been mentioned as a characteristic of the creative individual. With this general knowledge, it is possible to arrive at quick and accurate estimates of almost any quantitative question, which will then allow immediate analysis of new thoughts and ideas.

Creativity is a function of a number of components, of which knowledge, imagination, and suspension of belief are keys. A paradox regarding creativity is that in order to be creative, you must have pragmatically (noncreatively) defined the problem you are trying to address. After determining what the exact problem is, then the areas of knowledge and imagination can be combined to create original solutions. One of the other seemingly contradictory characteristics of creativity is that it works best when we do it over and over. By that, it is meant that we learn by doing.

As it is easy to see, there are specific qualities that creative people possess. However, it is important to realize that creativity is not a "you have it or you don't" proposition. It is a skill to be learned and practiced, just like riding a bike or juggling. By learning the techniques involved, it is possible for anyone of us to become one of those "creative" people.

Methods for Developing Creativity

Creativity is a habit, just like anything. The more a person does it, the better that person will become in generating original responses to various problems and challenges. However, we must understand that while individuals in many professions are encouraged to express their creative ability, and even get rewarded for doing so, most professions, including many engineering disciplines, are heavily weighted toward standard processes, conformity to the rules, and performing within a business structure.

In most western societies people seem to assign certain universal human abilities, like creativity, to only a fraction of all people, usually artists, musicians, architects,

> We don't need new technology to affect organizational performance, we simply need new thinking.
>
> — Gloria Gery, performance consultant

and so forth. This makes it more difficult for *all* members of the society to see themselves as creative people. This is a myth that we must be aware of and work to eliminate if we are to unleash our own creative abilities. One of the best ways to overcome this suppression of creativity is to actively work on bringing the innate creativity that lies within all of us, to the surface. As previously mentioned, it really helps to think of creativity as a skill or set of skills. By practicing these skills, one can get better at using them. There are many ways this can be done. Here are just a few of the methods that can be used to improve your creativity:

First of all, look beyond the immediate object or situation in front of you and see the "big picture." When the final portions of the transcontinental railroad were being finished, most people just saw tracks, trains, and a way to get across the country faster. But those who benefited the greatest financially were those who saw the railroad as a way to share ideas and products among more people. It was those same people who saw that the railroad would make it more likely that women, children, and the wealthy would be willing to travel westward, and that the land around the tracks would be where future cities would spring up. Don't just look at what's in front of you, look at the potential consequences.

Another way to improve your creativity is to always ask "Why?". We are surrounded by an unlimited potential of innovative ideas every day. They exist in the world that encompasses us. All we have to do is question the ordinary. Leonardo da Vinci was quoted as saying, "I roamed the countryside searching for answers to things I did not understand. Why does thunder last longer than that which causes it? How circles of water form around the spot, which has been struck by the stone. And how a bird suspends itself in the air. Questions like these engaged my thought throughout my life." Remember to continually question and always remain curious.

As often as possible, make time for relaxation activities and exercise to give the mind a rest and time for the subconscious to process information and make connections. Take time to let a solution define itself. An example of this would be the architect who, having finished a cluster of office buildings, was asked where the sidewalks should be placed between the structures. The architect told the construction crew to just go ahead and plant grass everywhere. A couple of months later, there were well-worn paths between the buildings that expressed the most efficient pedestrian traffic flow. The architect then had the sidewalks poured over these paths, which were a perfect response to the users' needs. Whenever you have a chance, try and create original ways to perform everyday activities. It will not only get you more used to expressing your creative abilities, but you will have more fun.

Also develop an interest in a variety of areas outside of your professional work. For example, browse libraries and read magazines and newspapers you normally do not, listen to various styles of music, and observe a wide array of artwork. Read books and literature, and attend courses on creative thinking techniques and practice them regularly. Many of these techniques are highly applicable to engineering design. Keep a daily journal as well and record your thoughts and ideas through the use of words and sketches. This allows the brain to take in new information and make connections with existing beliefs, thoughts, and emotions. A prevalent characteristic of creative people is that they are interested in a wide variety of subjects.

Lateral thinking is yet another way to improve your creativity. The definition of this term in the *Concise Oxford Dictionary* reads: "seeking to solve problems by unorthodox or apparently illogical methods." Lateral thinking refers to thinking from a number of different angles and perspectives to generate alternative solutions. Consider lateral thinking as the opposite of vertical thinking. Vertical thinking begins with a single concept and then proceeds with that concept until a solution is reached. As Edward de Bono states in his writings on lateral thinking, "Logic, or vertical thinking, is the tool that is used to dig holes deeper and bigger, to make them altogether better holes." But if the hole is in the wrong place to begin with, then no amount of improvement is going to put it in the right place. For example, when Fred Smith, founder of Federal Express, first proposed that instead of trying to determine all of the different methods and resources needed to deliver a package overnight from one location directly to another, he would bring them all to one place first, and then deliver them to their final destination. Instead of trying to improve the existing process (vertical thinking) as everyone else had always done, Mr. Smith questioned the whole philosophy and came up with a totally original idea. And as everyone well knows, it is a very successful lateral answer!

Mindmapping is a method that combines free word association and brainstorming. The idea is to generate as many ideas as possible by beginning with a central theme or topic. Using the theme or topic as a starting point, begin to write down thoughts that occur to you as quickly as possible, not analyzing anything. You will have words/ideas related to your central theme (main branches) and then words/ideas related to these branches (see **Figure 3.1**).

Figure 3.1 Example of the Mindmap Model of creativity

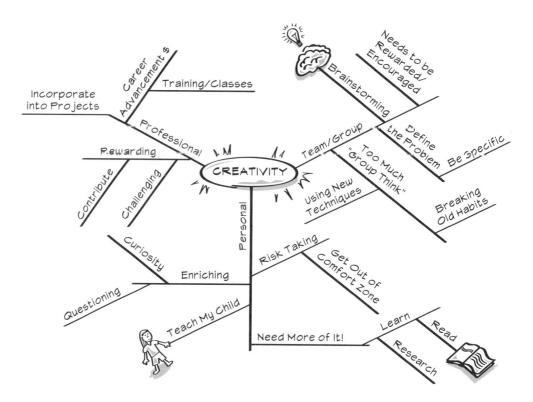

Brainstorming is a group problem-solving technique that involves the spontaneous contribution of ideas from all members of the group. The purpose of brainstorming is to get as many ideas as possible from a group of people in the shortest possible time. Quantity, and not quality, is the order of the day. It does not matter if the ideas are thought to be unworkable, or crazy, or outlandish by anybody within the group. Sometimes these ideas are the very ones that are adapted into other forms that solve the problem. For effective brainstorming to occur, the group must agree to not comment on ideas as they are presented, otherwise the brainstorming process becomes sidetracked. From a team standpoint, brainstorming represents an opportunity to receive input from all members of the team, not just the normally vocal ones.

Forcing new connections relates to connecting one thing with another to come up with a total new idea or a solution to a problem. Two popular methods for helping to force a new connection are: (1) observation — This method involves looking around you and making connections with things whether they show similarities or otherwise; and (2) random word — You use your finger for this method to point to any word in a dictionary, newspaper, magazine, or book. Then you relate whatever problem you have with this randomly chosen word. Choosing a word totally at random and making a connection with your problem forces you to be creative in arriving at the connection that may well be the solution to the problem!

Breaking the rules is another creative activity. One of the simplest ways of getting new ideas is to see what ideas have already been thought of, and then reversing or twisting these ideas to come up with new ones. The "breaking the rules" method allows for innovative solutions to common problems. For example, it used to be assumed that an elevator had to be put on the inside of a building, as it had always been done, until some creative architect thought about putting it on the outside of a building. This innovation created a more beautiful view to look at, provided a striking architectural feature, and saved interior building space, all at the same time.

The use of analogies (a comparison between two things) and metaphors (a figure of speech in which two different words or objects are linked by a similarity) can be used to equate a given problem and a similar problem that has a solution already defined. In many cases, designers use analogies to produce novel creative designs. These analogies are unique in that they often associate concepts that are not normally thought of as connected. That is, the concepts do not share any obvious features, but they do share underlying similarities and properties. As mentioned earlier, Leonardo da Vinci first thought of the concept of sound waves by observing the concentric ripples of water created by tossing a stone into a pond. Sir Isaac Newton's realization of the principles of gravity occurred to him after watching an apple fall from a tree. As he had done hundreds of times before, Archimedes watched the water rise as he sat in the bathtub one evening, only this time it dawned on him that the rising level of the water was tied to the relationship between volume and density. This realization occurred to him as he wrestled with the problem of how to determine, without damaging it, whether or not a wreath given to the king was truly made of solid gold. (Can you determine what his final creative solution was?)

Industry Scenario

A classic example of a simple analogy occurred during 1986, when renowned physicist Richard Feynman was part of the governmental commission that investigated the causes of the tragic explosion of the space shuttle *Challenger*. There was much debate concerning the performance of the rubber O-rings that acted as seals in the booster rockets. Feynman was able to prove that the explosion was due to the fact that the rubber O-rings lose their elastic properties at low temperatures. This had occurred to him while eating dinner and noticing a glass of ice water on the table. The next day Feynman illustrated his theory concerning the O-rings by placing a piece of identical O-ring material into a glass of ice water. After waiting a short time and upon pulling the O-ring back out and observing its inability to revert back to its original shape, the panel was provided an actual illustration of what had occurred on the shuttle *Challenger*.

A moment's insight is sometimes worth a life's experience.

—Oliver Wendell Holmes

Instead of just describing a problem using words, draw a picture of the problem. Many times, a visual representation of a problem will provide fresh perspectives and insights. It also helps in developing analogies and metaphors.

Look at something from a different viewpoint or perspective. Examine the needs, biases, and requirements. As an example, think of the architect who must view a project and/or problem from the view of the client, the builder, the suppliers, the financiers, and the community. By considering all of these perspectives, a better and more creative solution usually will be determined.

The Matrix/Checklist Method (see Figure 3.2) is a way to create new possibilities and it helps a person to ask lots of questions as well as consider lots of possibilities (a whole systems approach!). Before the concept matrix can be utilized, a list of ideas or possible solutions must be developed. After compiling the lists, the designer then places the ideas/possible solutions on a matrix with the vertical columns containing a list of one set of primary design characteristics and the horizontal rows containing a list with another set of primary design characteristics. After laying out the matrix, the designer now has a visual key by which to guide their decision-making process. The designer can look for combinations of design characteristics that have the possibility of making a new and unique solution. The key to success in this exercise is the quality of the primary design characteristics initially developed.

Figure 3.2 Example of a Matrix/Checklist Model

And of course, technology now provides us with creativity tools. There currently exists a multitude of software applications that will generate any number of possible solutions based upon information provided. Names of some popular idea-generation software products are: *Axon Idea Processor*™, *Brainstormer*©, *Genius Handbook*, *Thoughtpath*™, *Ideafisher*©, *Serious Creativity*™, and *Innovation Toolbox*©.

To be creative, you must have the mindset that nothing is set in stone. What is perceived as impossible is not necessarily the case. It may simply be that nobody had spent time thinking of alternatives because what's currently being done seems to be working well enough. For instance, during the oil crisis in the 1970s, solar power and other tools tapping nature's energy quickly emerged. However, after the crisis was over, it was back to the comfort zone of cheap and easy oil. Development of alternative sources of energy faded into the background. But the fact that alternative sources of energy can and were found shows that we have the ability to discover what is not obvious, if only we allow ourselves to envision it.

Again, creativity is a habit and a practice to be learned and developed. Information regarding creativity can be found from numerous sources to help a person increase their own creative abilities. These include books, software products, the Internet, and of course, other people. And since we learn best from observing and doing, try to spend time with people who you believe are creative and ask them how they developed their creative skills. Start by being more creative in your everyday life and thinking about unique solutions to common tasks. Be curious and keep asking questions!

TRIZ: A New Approach to Innovative Engineering and Problem Solving

Developing new and innovative products and processes in today's fiercely competitive global marketplace is an immense challenge. Companies must meet that challenge with limited resources, higher quality, lower costs and quicker time-to-market products.

Conventional approaches simply will not get the job done. Techniques for solving routine technical problems such as the use of CAD/CAM/CAE, finite element analysis, and 3D computer simulation substantially facilitate development of the existing concepts, but do not generate breakthrough concepts. Systematic innovation in products and processes is critical for a sustainable competitive advantage, but it is possible only if the approach to idea generation is equally unconventional. The TRIZ process allows for just such an approach.

TRIZ (an acronym derived from the original Russian term for the process) was developed in the former Soviet Union by scientist and inventor Genrich Altshuller. Altshuller's work with TRIZ began in the 1940s. Since that time, there has been much interest and application of the TRIZ concept to various areas of human activity.

TRIZ research began with the hypothesis that there are universal principles within the invention process that are the basis for creative innovations, and that if these principles could be identified, they could be taught to people to make the process of invention more predictable. After reviewing over two million patents, classifying them by level of inventiveness, and analyzing them to look for principles of innovation, three common threads emerged. They are as follows:

- Problems and solutions were repeated across industries and sciences.
- Patterns of technical evolution were repeated across industries and sciences.
- Innovations used scientific concepts outside the field where they were developed.

This patent research indicated that only two percent of solutions were truly pioneering inventions; the rest presented the use of a previously known idea or a concept in a novel way. Thus the conclusion was that an idea of a solution for a new problem might be already known.

It was also found that common patterns exist not only between individual solutions. Similar patterns were observed between the development of different technical systems and design products over time. A further conclusion was made that the evolution of technology is regular, not random.

Modern TRIZ should not only be seen as a methodology for solving engineering problems or a new product development process, but as a powerful tool for managing knowledge and solving problems that contain contradictions in many areas. As reported, many successful results were obtained in solving social, business, marketing, and managerial problems. Large and small companies around the world use TRIZ for many applications to solve real, practical everyday problems and to develop innovative strategies for the future.

A case study using TRIZ: NASA and the Venus probe

Initial Situation

In the 1990s, the United States' NASA space agency was designing an autonomous probe to land on Venus. The probe had to carry various electronic devices to the planet. When the project was close to completion, the agency got a request from a group of scientists, headed by a renowned chemist, to place one more device into the probe. This was impossible to do, for the probe was already so crammed with other devices that one could hardly tuck a matchbox in between them, let alone the 13-pound device being requested. A creative solution was needed to solve the problem . . . How to find extra space for the device?

TRIZ Analysis

As it often happens, a seemingly impossible problem may have a very elegant and belatedly obvious engineering solution. One approach commonly discovered in the TRIZ research calls for delegating several functions to one object. So, for NASA the question became, What other existing device can provide the required function? The only way to "squeeze in" the extra device without removing another one was to integrate functions of the former with some already existing resource.

Solution

A detailed analysis of the probe design revealed the previously overlooked opportunity. Each planetary probe built earlier had carried to outer space almost 13 pounds (what a coincidence!) of a balancing weight made of cast iron. This balancing weight controlled the position of the probe's center of gravity during landing. The solution was to replace the balancing weight with the proposed device that could perform both functions — its scientific duty and positioning of the center of gravity.

As can be seen in this example, TRIZ saves time in idea generation, also called ideation, by focusing on specific solutions or techniques that have been effective in other design problems. Most importantly, TRIZ unlocks the grip of psychological inertia that blocks engineers, designers, and technicians from finding ideal solutions. When psychological inertia is broken, unique, novel, and truly innovative solutions are frequently discovered.

TRIZ approaches can find strategic opportunities for their teams and organizations that other models are not able to identify. They can capitalize on these opportunities by developing new products and processes as well as services or organizational structures. And the ability to see opportunities before they are obvious can provide a huge competitive advantage.

Individual Creativity versus Group Creativity

By its very nature, the creative process involves the integration of ideas, bouncing options back and forth, thinking beyond oneself. Creativity and innovation are, in many cases, a team effort. One person can come up with new ideas and generate pieces of the solution, but the creative process is enriched when there is the benefit of multiple input and varied styles of thinking. When a team works together pooling their creative ideas, more options are generated to meet the challenges of the team.

Figure 3.3
The creative process is enriched when ideas are pooled

Teams, or groups are generally superior to individuals in generating creative solutions. This is due to several facts:

- A group is able to provide the varied perspectives and accumulated knowledge and experiences that are needed for creativity.
- Groups are better able to minimize or eliminate errors and mistakes.
- Individuals within a group will learn from one another, which provides a self-perpetuating synergy, and consequently, more creativity.

However, there is a negative aspect as well. Group creativity must struggle with the reality that most people are willing to go along with the group, and that individuals are sometimes very shy in offering up ideas that may initially appear unpopular or unrealistic. This can lead to compromise and ultimately, mediocrity. It is therefore extremely important to educate groups, or for the group to have in their norms a requirement that all voices must be heard. Creative groups will tolerate more ambiguity, more risk-taking, more willingness to offer up and consider "over-the-edge" ideas. Truly creative groups are willing to give up individual control and authority. If this can be accomplished, then it has proven that as the number of individuals in a group increases arithmetically, the value of the creative process increases exponentially. Adding a few more members can dramatically increase the value of the group for all members.

Creativity and innovation requires a diverse, information- and interaction-rich environment. It requires people with different perspectives working together toward a common goal, who possess accurate, up-to-date information and the proper tools. Innovation also requires a learning environment. Only by creating such an environment where every member of the organization is continuously learning more about its products, services, processes, customers, technologies, industry and environment, can an organization successfully innovate year after year. One of the most significant ways that business is creating this continuously learning organization is through the use of what is known as knowledge management. Knowledge management is an attempt to accumulate the collective wisdom of an organization, allow its employees to find any portion of this information in a timely manner, understand it, and use it effectively. This is usually accomplished through the incorporation of a large computer database that is networked to all of the business's employees. It also contains a powerful search engine and a method for individuals to input information into the database in a logical manner. Knowledge management has become critical to an organization's ability to harness the best practices and full potential of its employees. As Margaret Wheatley states so eloquently in her book, *Leadership and the New Science:*

> We do not exist at the whim of random information . . . Our own consciousness plays a crucial role. We, alone and in groups, serve as gatekeepers, dreading which fluctuations to pay attention to, which to

Knowledge Management — a tool for creativity and innovation.

suppress. We are already highly skilled at this, but the gatekeeping criteria needs revision. We need to open the gates to more information, in more places, and to seek out information that is ambiguous, complex, of no immediate value. I know of one organization that thinks of information as salmon. If its organization streams are well stocked, the belief goes, information will find its way to where it needs to be. The organization's job is to keep the streams clear so that the salmon have an easy time of it. The result is a harvest of new ideas and projects.

And finally, one of the most important keys to individual or group creativity is clarity. Never work on a creative challenge without first clearly defining the problem in the form of a question. If your work involves a team, post the question throughout the office where everyone can see it. Then get people to agree on the words you use. For instance, if your creative goal is to improve a product, you have to first be sure that everyone has the same definition of "improvement."

Creativity versus Innovation

It has long been recognized that both creativity and innovation are required to bring about the results necessary for obtaining a competitive advantage in business. As business management authority Peter Drucker puts it, "Business has only two functions, marketing and innovation." Those companies that are most successful in rapidly developing innovative new products and services will have the required competitive edge in today's world. The question that must be constantly asked then is, How can we do it differently? This is the continuous challenge for the engineering and technical community.

The creative process results in innovation. Innovation is the transformation of an idea into a tangible product or service. Innovation usually occurs because one of two things has happened. Either an original idea is developed from a new technology (technology push), or a new demand in the market creates the need for a new product or service (market pull). While creativity is the ability to make or bring a new concept or idea into existence, innovation is the process of refining creative ideas to do something in a new way: to affect a change to create a new idea, method, or product. The distinction between creativity and innovation is important. An individual business, industry, or society might develop an unexpected and revolutionary idea for building a better product, but true product innovation will occur only when the creative idea is transformed into something tangible, which will then provide the business, the market, or the society with a unique and more effective alternative. A powerful and defining example of this occurred when Europe (the innovator) harnessed the power of competitive innovation in a way that China (the creator) did not. Europeans used the Chinese invention of gunpowder in the eleventh century by creating various weaponry which used gunpowder, such as guns and cannons, and then used these innovations to eventually rule the New World for centuries.

Whereas creativity (creation) is commonly an individual process, innovation is almost exclusively a group process. This is because individuals rarely possess all of the knowledge needed to transform a creative concept into an innovative solution. This transformation of an idea or concept requires an analysis by any number of different individuals or groups. These may include manufacturing, marketing/sales, finance, customer service, or any other group within the business organization that is potentially impacted by the original idea. In other words, innovation is normally only successful by applying a whole systems approach. Without a systemic approach, creating solutions in one area can often cause problems in another. That is not innovation, it's chaos.

Innovation is usually divided into two basic types, incremental and radical. Incremental innovation occurs when an existing product, process, or service is improved upon. Radical innovation is when an entirely new approach or idea is used to create an original product, process, or service. Even though radical innovation occurs in a much smaller percentage than incremental innovation, the benefit from radical innovation considerably outweighs that of incremental innovation. Each type performs a necessary role in providing society with more and better choices (see Figure 3.4).

> CREATIVITY is thinking up new things. INNOVATION is doing new things
>
> —Theodore Levitt

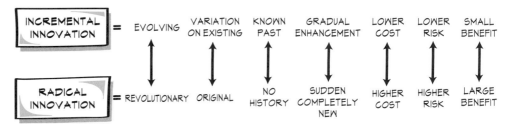

Figure 3.4 Comparison of information versus radical innovation

Businesses are much like humans in that they are predisposed to strongly defend what they know and what they think works. These businesses find changing both unnecessary and, in many cases, impossible. There is simply no room in most companies for the concept of giving up a process, service, or product that is working. This natural tendency is what is leading many companies today to fall into desperate situations before true change is seriously considered. Truly successful companies embrace change, creativity, and innovation as part of their philosophy of doing business. This philosophy is ingrained in such companies as Xerox, Walt Disney Studios, 3M Company, and Apple Computer. Today's corporations must consistently be willing to let go of past successes and embrace innovation, because those successes will be replaced by innovation from others. To operate in this type of environment, successful companies develop, nurture, and promote an innovative climate in the workplace. Some of the ways in which this is done follow.

- Discover how individuals are most creative: Ask for their input, keep in mind that people have different work styles.
- Define challenges specifically: Focus on areas where creative solutions are needed.
- Minimize fear of failure: Find ways to absorb risk. Regard mistakes as learning opportunities.
- Embrace play: Give people opportunities to master play skills.
- Encourage active communications: Support ongoing interactive exchanges of ideas.
- Allow adequate time for ideas to develop and mature.
- Define creativity as a fundamental aspect of the company's success. Tie creativity to specific business goals.

Innovation through Failure

Failing *can* be a learning experience and a necessary step in the design process. Failure also teaches you much more than success and many times it is actually *unrealized* success. By this, we mean that what at first appears as a failure subsequently may become an ideal solution to an entirely different problem.

The ability to recognize other applications for failed experiments has provided amazing success in a variety of products and product development. Several illustrations will help explain. A number of years ago, an inventor named James Wright was trying to invent a rubber substitute made from silicon. One of the results of his efforts was a gooey material that stretched and bounced, but did not appear to have any practical use. About five years later, the idea occurred to put portions of this material in plastic eggs and sell it as a children's play toy. We now know this as Silly Putty. Another example of a "successful" failure is the 3M Post-it® Note. Originally a failure in the effort to create a stronger adhesive for tape, a 3M Company researcher realized that the weaker adhesive would make a perfect bookmark, one that would initially stick, but could be removed without damaging the book. And as the saying goes, the rest is history. The applications of 3M Post-it® Notes were greatly expanded and are now one of the primary ways information is communicated and organized in both our business and personal lives.

> Remember the two benefits of failure. First, if you do fail, you learn what doesn't work; and second, the failure gives you the opportunity to try a new approach.
>
> — Roger von Oech

History is filled with examples of situations that were originally considered mistakes, but subsequently turned into important discoveries. Before creating the light bulb, Thomas Edison discovered 1,800 ways not to build one. One of Madam Curie's failures was radium. Christopher Columbus was looking for India. Errors are one of life's primary methods of learning.

Finally, remember to look at every failed attempt to do something not as failure, but as another piece of experience and knowledge that will allow for future success. It is old wisdom that states that the only people who have never failed are those that have never attempted to succeed.

> ### Industry Scenario
>
> **Xerox PARC: Innovation and Creativity in Practice**
>
> In 1970, Xerox Corporation gathered a team of world-class researchers and gave them the mission of creating "the architecture of information." The scientists of the Palo Alto Research Center (PARC) lived up to this challenge by inventing personal distributed computing, graphical user interfaces, the first commercial mouse, bitmapped displays, Ethernet, client/server architecture, object-oriented programming, laser printing, flat panel displays, and many of the basic protocols of the Internet. In the years since Xerox established the Palo Alto Research Center, PARC technologies have changed the world. The Xerox PARC philosophy is that there exists something more than just technical expertise or intellectual brilliance. When hiring individuals to work at PARC, they look for attributes such as intuition, risk taking, pattern recognition, and the ability to understand links, and the desire to listen and learn. Xerox PARC believes that to do things differently, we must see things differently. The great challenge in innovation is the ability to link emerging technologies with emerging markets. Xerox PARC realizes that innovation is the product of creative tension between differing viewpoints, and that you must be willing to give up cherished beliefs and see from new perspectives. For living this philosophy, Xerox PARC is recognized as one of the premiere research centers in the world.

Innovative ideas are the cornerstone for success in any business or engineering environment, be it public or private. Kevin Kelly, the editor of *Wired* magazine, makes the following statement regarding the new rules of the global economy: "The new rules governing this global restructuring revolve around several axes. First, wealth in this new regime flows directly from innovation, not optimization; that is, wealth is not gained by perfecting the known, but by imperfectly seizing the unknown. Second, the ideal environment for cultivating the unknown is to nurture the supreme agility and nimbleness of networks. Third, the domestication of the unknown inevitably means abandoning the highly successful known — undoing the perfected." As is readily apparent, our ability to solve our social and economic problems will be limited primarily by our lack of imagination in seizing opportunities.

Creativity, Innovation and the Engineering Design Process

As with any engineering design concept and process, the ultimate goal is to provide an improved product, process, or service. In addition, it is imperative that this goal be accomplished in the most cost-effective way possible, through shortened development time, improved production or construction methods, or more efficient business practices. The creative and innovative process is normally emphasized in the initial phases of the engineering design process. However, creativity and innovation can and should be used throughout the engineering design process to develop better design processes, testing and evaluation methods, criteria definition, and investigative techniques (see **Figure 3.5**).

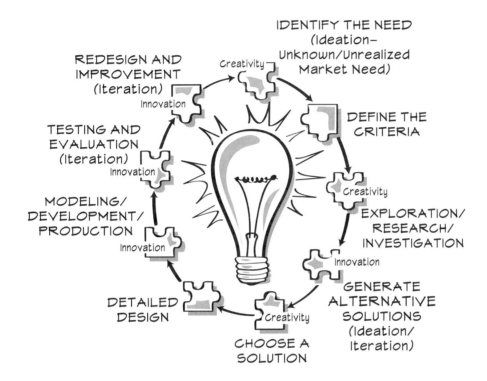

Figure 3.5
Creativity, innovation, and the engineering design process

Creative thinking must be combined with analytical thinking, since they are not mutually exclusive. To be truly innovative, you need to do both. Analytical thinking involves discussing the realities of a situation or problem and determining which course of action will best meet the inherent challenges. Traditionally, analysis has been the foundation of the engineering process. The key to competing in today's market is for the engineering organization to use its extensive analytical skills and apply them creatively to develop innovative products. Refining and selecting creative ideas and building plans of action are just as important to innovation as the creative process. Creativity and innovation efforts require specific goals and parameters to follow.

One of the items that is crucial in any innovative effort is to define the customer's/client's needs and commitment to the process. The following are some questions that should be answered prior to beginning a new project.

- Is the customer/client clear on what they want?

- Is the customer/client willing to pay the price in increased uncertainty, failures, perceived lack of control, and in resources for what they want? These are vital questions because companies often say they want to become more innovative, but are not prepared to tolerate the associated increased risk.

- Do they understand the necessity of increasing effective communication between all parties involved in the process?
- Make sure that the customer/client understands that if there is more risk associated with the project, there will be a greater degree of creativity and innovation required. Clients must recognize that operating in this environment may be uncomfortable, but it is necessary to a successful outcome.

Let's examine the case of a company who has used innovation as the key to their success. When Autodesk, Inc., the current sales leader in CAD software applications, initially developed an application called AutoCAD in the early 1980s, it needed a way to penetrate the computer-aided drafting/design market. It came up with an innovative solution. It decided to provide its product to educational institutions at an extremely low price. Since most schools operate on a very limited budget, these schools jumped at the opportunity. This strategy had the ultimate effect of training large numbers of future CAD operators on their product. Autodesk realized that eventually, these would be the very people making buying decisions for CAD software. Autodesk thus obtained the name recognition and market share it desired. Only after the company had reached its position as a market leader, did it begin to increase its price to more accurately reflect the products' development costs. This innovative solution has allowed Autodesk and the AutoCAD product line to be one of the most dominant in the CAD/CAM industry.

Innovation and the Use of Rapid Prototyping

The quest for faster speed-to-market has become a design imperative. Increasingly, organizations are using computer-aided design and engineering systems to accelerate product development. Service companies also are using digital systems to increase their rate of innovation. Speed is significant and fairly straightforward to measure, but speed-to-market metrics must be handled with great care. An accelerated rate of innovation can outstrip the customer's ability to absorb it.

"Fail early and fail often" is the philosophy behind successful prototyping. Stefan Thomke of the Harvard Business School persuasively argues that organizations which aggressively use models to identify design conflicts early in the development process achieve significant economies of innovation over firms that do not. This approach, which Thomke calls "frontloading" can, he asserts, "reduce development time and cost and thus free up resources to be more innovative in the marketplace."

In effect, rapid prototyping means rapid failures. This is no mere paradox but a vital design principle that world-class companies are embracing. Digital modeling encourages the "frontloading" of failures. Industry leaders such as Toyota, Boeing, Ford

Motor Co. (see color insert), and DaimlerChrysler have become adept at generating simulations that offer accelerated insight into key design conflicts. Frontloading captures these critical conflicts as early in the development cycle as possible. Given that otherwise these design conflicts have to be corrected later in the process when they are more expensive to fix, this "fail-early" approach is becoming an imperative.

Thomke's work with German automaker BMW revealed that computer simulation of crashworthiness simultaneously accelerated the pace of design iterations and reduced their costs. Reviews of ninety-one design iterations showed that developers were able to improve the side-impact crashworthiness of a BMW sedan by about 30 percent — "an accomplishment that would have been unlikely with a few lengthy problem-solving cycles using physical prototypes only," according to Thomke.

The same "design-for-failure" ethic is applicable to the simulation of experiences. Financial-service firms that create and sell high-risk financial instruments are increasingly investing computer time and human creativity in scenarios and simulations in which their instruments fail. Given the volatility of global markets, these stress tests enhance the credibility, and salability, of the innovative instruments that survive. The key point is that financial innovators, like their engineering counterparts, use digital technologies to frontload their failures, so that those failures never reach the customer.

You know you have a successful prototype when people who see it make useful suggestions about how it can be improved. Successful innovators don't use their prototypes to persuade the right people; their prototypes enable the right people to persuade themselves. If your most important boss, client, or supplier can't play creatively with your prototype, you have made a serious design error. Prototypes should lure people into innovative games of "What-if?" and turn customers, clients, colleagues, and vendors into collaborators.

As cross-functional, cross-disciplinary teams become a dominant medium for managing innovation, prototypes and simulations can promote awareness and understanding between collaborators. It is a useful exercise for a design engineer to simulate the constraints imposed on manufacturing engineers and for an insurance agent to experience those confronting the claims adjuster. It is as valuable for an airline pilot to simulate the task of an air-traffic controller as it is for the air-traffic controller to sit in the cockpit of a Boeing 737 simulator. Conversations during such role-playing tend to be insightful. It is not enough for prototypes to be the shared medium of communication and collaboration in the innovation process; they should also be a tool for people to step outside of their everyday roles. Why? Because seeing the value of a prototype or simulation requires viewing it from a different perspective.

Summary... Links to the Whole

As the world of engineering continues to develop into an environment that is driven by continuously evolving market desires and needs, the ability to be creative and innovative is crucial to everyone. As illustrated in this chapter, companies are placing a premium on the skills involved in creativity. Business is now designing questionnaires and interviews to determine if potential employees possess the characteristics necessary to promote creativity and innovation. Creativity is no longer just an overstated theme mentioned by every company in its mission statement, but a required component of staying ahead in today's business world. To achieve this, individuals must practice the individual habit of creativity and learn to operate within the team/group environment. The skills of creativity and innovation are encompassed in the whole systems approach to the engineering design process and the holistic knowledge required of the individual. In our everchanging, information-based society, the only companies that will rise to the top are those that can access and fully utilize their most valuable resource — their people.

Creativity is a life skill that must be continually cultivated. Remember, being creative is *looking* at the same thing as everybody else but *seeing* something different.

A-Team Scenario

The A-Team Brainstorms the Best Way to Meet the Pump Design Requirements

"So, where do we start, folks? Joe exclaimed, "We have a pump to design." "Where do we start and how do we proceed?" Shantel followed with a tone of ambivalence. "How do we organize what we have to do?" "Well, don't we have to consider all the design specifications that Earl wanted for this pump?" Carlos added. "Yep, you are right on and that's where we will start. Let's start with just one concept and explore that before we discuss any of the others," Joe suggested. "Which specification do we look at first?" Carlos asked. "Well," Claudia jumped in, "we know he wants it small and it has to fit on a ranch or farm vehicle and it has to be able to be functional in different directions and it . . ." "Whoa! Wait up, Claudia. Let's look at just one of those specs," Carlos exclaimed. "It seems to me if we knew where a pump

TOOLBOX

Creativity and Innovation in Design

For Chapter 3, the tools consist of:
- Idea generation matrix
- Mindmapping
- Brainstorming
- Lateral thinking
- TRIZ problem-solving method
- Rapid prototyping

A-Team Scenario
continued

could be attached to a vehicle, we would then also solve how many orientations are really needed." "Carlos is right!" said Joe. "Let's brainstorm ways a pump could be attached to a vehicle. Any ideas on how we can figure this out?"

The group went on to creatively brainstorm ways to place a pump on multiple types of vehicles. Then Claudia showed the group how to utilize some of the brainstorming ideas in a mindmapping exercise. This allowed for creative input from all of the team members. Shantel reminded the A-Team of their norm that states that there are no dumb ideas and that everyone has to contribute.

Once they got a fairly long list of potential placement locations on the vehicles, they started sorting them into orientations. What they discovered is most could be oriented in a top-to-bottom position or sideways position. This would save costs by simplifying the design. As they all sat and studied their lists, Carlos remembered that one of the first suggestions for developing creative solutions is to approach the problem from a different perspective. So he jumped up out of his chair and laid down on the floor so that he was under the table looking straight up at the bottom of the table. Everyone started yelling, "What are you doing?" Carlos said, "Servicing our pump design. If they place it in the top-to-bottom orientation under a vehicle, they will have to design a hanger part to mount it and that makes it difficult to service or remove and costly to assemble." Joe is thinking now, "Carlos has a point. Maybe if the pump had a mechanism in it that allowed it to swing down or up it could still be placed in that orientation. That would save additional design work for the pump for multiple orientations. Russell questioned, "If the pump were able to swing, does that mean we need more clearance space for the pump location?" Shantel pipes up, "Guess we will have to research that, but we only have six months to get this done. It seems we have a lot to figure out."

"Hey, we have to think smart here and use the technology tools we have at our disposal," adds Carlos. "We can answer some of these questions by building a solid model in the computer and simulate different positions and clearance spaces." Claudia offers to go on-line and check out truck and farm vehicle Web sites and request some information about their design specifications.

This brainstorming process the group used was very helpful in thinking through one of the specifications. What became apparent was that all the specifications were related to each other. Once they knew the possible orientations and positions in which a pump could function, they had to determine what type of materials to use for the pump. The orientation of the pump was one factor to consider because of material weight but now the group realized that the application or what the pump was pumping also determined the materials to be used. The brainstorming process started all over again.

Respond to the A-Team scenario by answering the following questions. Use any tools or ideas you have learned in this chapter.

1. List here the creativity tools that the A-Team utilized with their pump project.

    _____
    _____
    _____
    _____
    _____
    _____

Design Tools for Engineering Teams

A-Team Scenario
continued

2. What other creative tool could the team implement in this discussion?
   _____
   _____
   _____
   _____
   _____

3. Did the A-Team move from creativity thinking to innovative thinking? Explain.
   _____
   _____
   _____
   _____
   _____
   _____
   _____

4. Discuss the group dynamics you observed in the A-Team's positive and negative aspects of the creative process for generating ideas.
   _____
   _____
   _____
   _____
   _____
   _____
   _____
   _____
   _____

TERMINOLOGY & DEFINITIONS

brainstorming — a group problem-solving technique that involves the spontaneous contribution of ideas from all members of the group

creativity — the ability to make or bring a new concept or idea into existence; marked by the ability or power to create

ideation — the forming of ideas or concepts

innovation — to do something in a new way; to affect a change; a new idea, method, or device

TERMINOLOGY & DEFINITIONS

iteration — the action or procedure in which repetition of a process (such as the creative process) yields successfully better results

lateral thinking — a process of seeking to solve problems by unorthodox or apparently illogical methods

matrix — a rectangular array of elements arranged in rows and columns that can be combined to produce various sums or solutions.

mindmapping — a method that combines free word association and brainstorming. The idea of mindmapping is to generate as many ideas as possible by beginning with a central theme or topic.

Activities and Projects

- Select a project related to your specific design discipline (architectural, civil, mechanical, electrical, etc.). Using one of the creativity methods described in the chapter, along with your own natural creative abilities, develop an innovative aspect to a specific part of the design project. Document your ideas, method, rationale, and desired outcome for this innovation.

- As one of the recognized "creative" experts at your product development company, you have been assigned the task of generating at least twenty original ideas for a new drinking cup. Work in groups of three to five with your fellow creative experts to generate the list of ideas. Hint: Setting up some specific criteria to indicate whether or not you are reaching your goal will be tremendously helpful, such as:

  1. Ideas that are aesthetically unique
  2. Ideas that are practical or functional
  3. Ideas that are aimed at a specific audience (i.e., children, golfers, or hikers)

- Generate "back of the envelope" quantitative answers to the following questions using both your creative abilities and estimating skills.

  1. How much snow (in cubic feet) sits on Mount Rainier in the summertime?
  2. How many golf balls would it take to fill a refrigerator?
  3. How many grains of rice are there in a 25-pound bag?
  4. How much is a company paying (in dollars) per reader when it chooses to run a full-page advertisement in the *Wall Street Journal*?

  Document your process, assumptions, and calculations. Compare your estimates with the class. Research and/or calculate actual answers to evaluate your estimates.

Activities and Projects *continued*

- The company you work for has determined that the contemporary model of an office environment is not conducive to creating an enthusiastic, innovative, and functional design team. So it was determined that the best way to create such an environment was to let the design team work together to arrive at an innovative solution. The criteria that has been established is that every individual's office should provide a clear statement of personal territory and a sense of privacy. But, at the same time, it must accomplish this without cutting everyone off from one another, the way office cubicles traditionally do.

    Divide the class into design teams of three to five people and have each team develop layout sketches, material lists, and a rationale for a minimum of two design layouts of an innovative work/office area. The layouts should accommodate personal workspaces for the team members, a space for team meetings, and a common equipment area (i.e., copier, fax, paper supply). Have each team present their preliminary designs, with the other teams providing constructive feedback through the use of questions, comments, and/or suggestions. Let each team use this feedback to reevaluate their designs and make any modifications or improvements they feel are valuable.

Landing Gear: Evolution of Design

As described in Chapter 1, An Introduction to Whole Systems Thinking, the engineering design process is a sequential series of steps or actions that a designer takes when solving a problem or creating a new design. In the following illustration, we can visually trace the design evolution of a specific product. In this example, the product is a retractable front landing gear assembly for a popular business jet airplane.

Identifying the need

Upon identifying a specific need, in this case, a new type of retractable front landing gear, discussions and ideas are generated to produce a concept that addresses an initial solution to the identified need. Often, these initial solutions are created in the form of sketches, drawings, or other visual representations. Shown here is the concept for a new retractable front landing gear that was initially just sketched out on a napkin.

(Courtesy Autodesk)

Design Tools for Engineering Teams

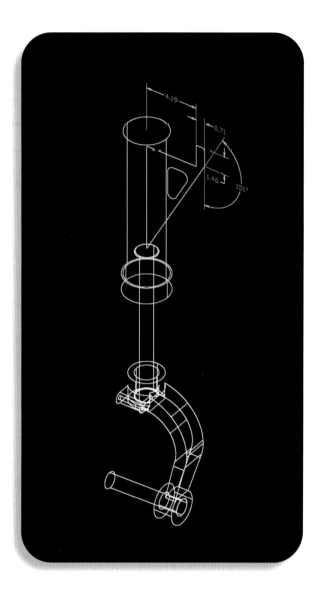

Defining the performance criteria and research

In determining and selecting the product's performance criteria, designers often utilize wireframe modeling applications. Using various software design tools allows the designer to manipulate various characteristics to achieve an optimal design, while providing the ability to visually assess the product's dimensional and aesthetic properties. During this phase of the design process, research is done to determine basic functional requirements, appropriate materials and manufacturing methods and processes, and establish projections for cost, production timelines, labor allocation, and vendor contracts.

(Courtesy Autodesk)

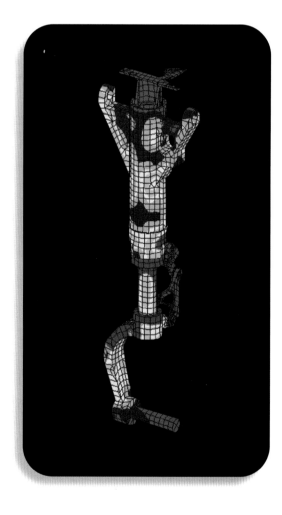

Finalizing the design

After generating a number of alternative designs for consideration and review, a final design is selected. As this illustration shows, a computer model of the design is then developed which contains information about the product's specific components (including purchased parts), dimensional clearances, range of motion, and interaction with mating parts.

(Courtesy Autodesk)

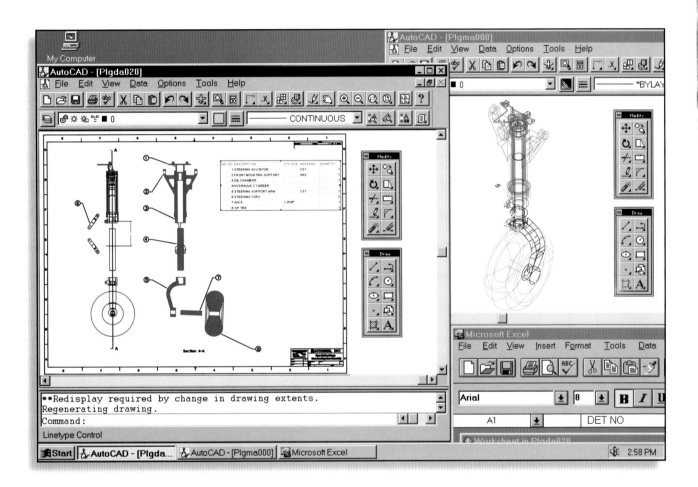

The detailed design drawings

After the design is finalized, it is necessary to develop all of the detailed engineering drawings that will be used to document, manufacture, purchase, and assemble the product. With current CAD software applications, it is possible to extract information from the finished 3D solid/wireframe model to quickly produce the two-dimensional detail drawings necessary for production, such as this image.

(Courtesy Autodesk)

Design Tools for Engineering Teams

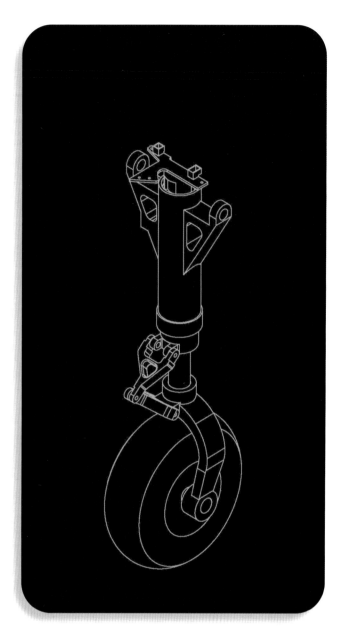

Modeling/product development

As the product begins to take shape in the development and production (or construction) process, the design teams need to make sure that the product will be able to meet the performance criteria, specifications, and goals established in the initial phase of the design process. This is accomplished through the use of detailed computer models (images shown), prototypes, computer-based simulations, and limited production of the product.

(Courtesy Autodesk)

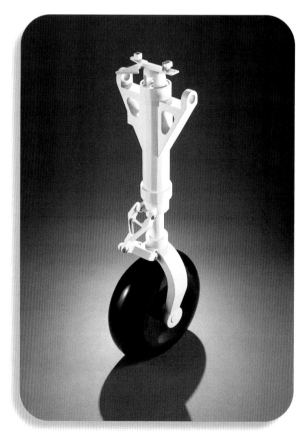

Design testing and evaluation

During the testing and evaluation phase, the goal is to see if the product will truly meet goals and expectations when placed under actual operating conditions. One method of testing that has become a critical step in the design and development process is that of computer-based simulations. Utilizing fully rendered solid models, the computer can perform the tasks of assembly and maintenance, simulate motion to determine clearances and interferences, and test for structural integrity under predetermined stresses and strains. This evaluation and testing tool has allowed for faster design cycles, lower costs, and improved reliability. This image is a computer-rendered model showing the front landing gear located relative to the landing gear doors and fuselage to determine if clearances are correct.

(Courtesy Autodesk)

Design Tools for Engineering Teams

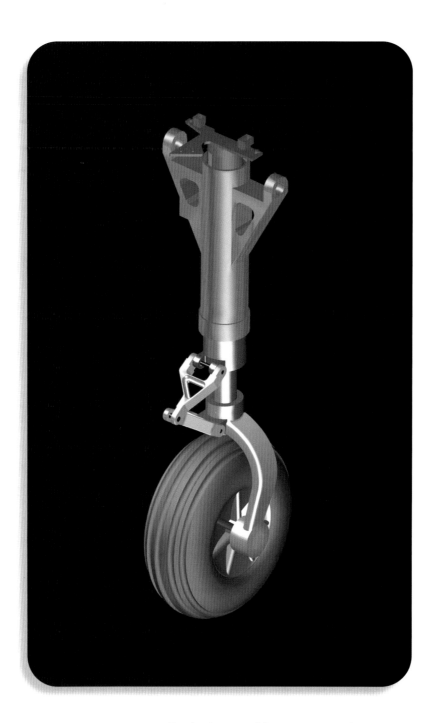

Redesign and improvement

One of the truly revolutionary changes in product design with the use of computer models is the ability to play the "what if?" game. Because of the ease in changing design parameters utilizing software design tools, design characteristics can be constantly modified in an attempt to improve a design. As this image shows, it is possible to experiment with almost any factor in the design, including shape, part location, material, part dimensions, or even colors.

(Courtesy Autodesk)

(© 2001 Shurgard Storage Centers, Inc. All rights reserved.)

An Architectural Project: The Development Process

The following illustrations show samples from the development sequence of a commercial retail facility in Seattle, Washington. This structure was built in a transitional zone between an industrial region and a residential neighborhood. The goal was to blend in subtle design touches that included interesting lines carved into the façade, sharply pitched roofs suggestive of residential homes, and strategically placed trees and shrubs, while at the same time architecturally incorporating the area's industrial history. The following comments will help to illustrate the engineering design process as it applies to the architectural/construction engineering discipline. Note the changes that are made throughout the engineering cycle that come about as a result of nonengineering causes. These are frequently system-related impacts and not engineering-related impacts.

Design Tools for Engineering Teams

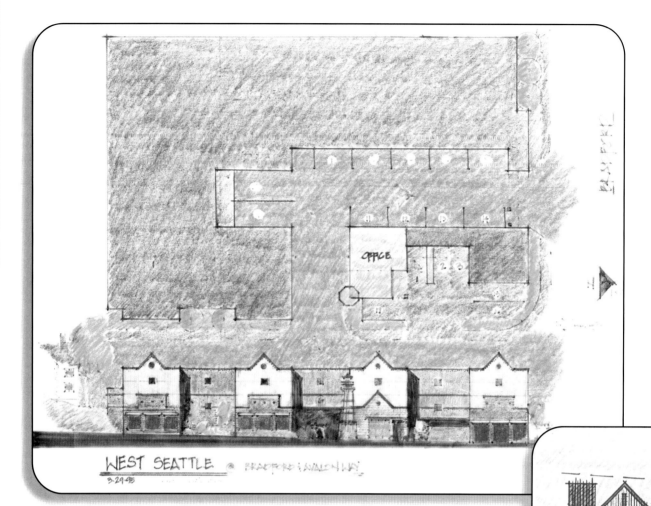

(© 2001 Shurgard Storage Centers, Inc. All rights reserved.)

Initial architect concept

Preliminary conceptual sketch showing an initial floor plan and the exterior elevation from the building's east side. This initial concept by the architect shows four raised and peaked exterior facades along with the company's trademark lighthouse. The lighthouse, envisioned by the architect in this instance, is a metal framework structure that symbolizes the industrial businesses that are an integral part of the neighborhood (there is a steel mill directly across the street). The upper portions of the peaked facades are wood frame and painted in a sharply contrasting shade of beige.

(Drawings and images courtesy of Shurgard Storage Centers Inc.)

Architectural rendering

An architectural rendering of the same east elevation drawn five months after the initial conceptual sketch now shows two major changes. The upper half of the raised peaked facades are now shown painted the same shade of brown as the rest of the exterior to limit the contrast and provide a more singular look to the structure. The other change is the revision of the lighthouse structure from a metal framework to the more traditional wood frame and glass materials commonly used by the company. This change occurred due to a management decision to maintain the company design standards.

(Drawings and images courtesy of Shurgard Storage Centers Inc.)

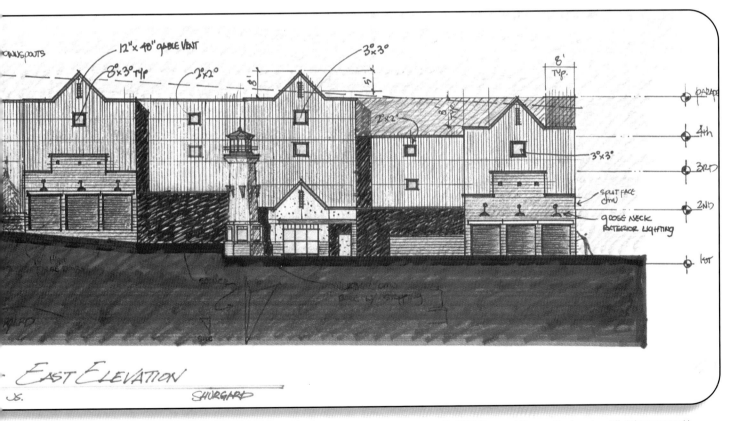

(© 2001 Shurgard Storage Centers, Inc. All rights reserved.)

Design Tools for Engineering Teams

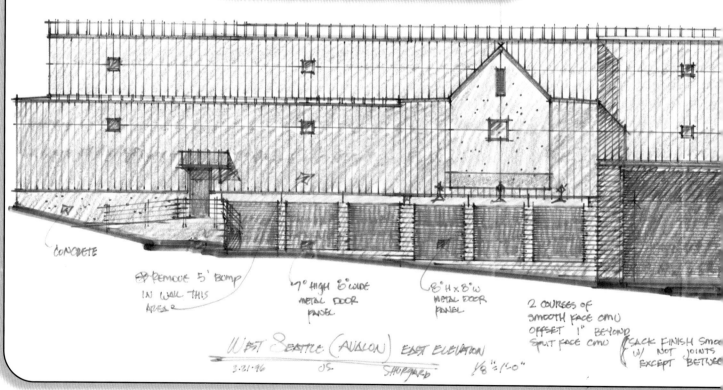

(© 2001 Shurgard Storage Centers, Inc. All rights reserved.)

Subsequent architectural rendering

This architectural rendering was drawn up some seven months later. It shows more detail along with some significant changes. Local zoning height requirements have necessitated that the peaked facades be lowered so that they now fall below the roof line. Due to construction cost concerns, the original four peaked facades have now been reduced to three, and the size and material of the facades have been changed to provide a more architecturally pleasing appearance, while creating more significant visibility.

(Drawings and images courtesy of Shurgard Storage Centers Inc.)

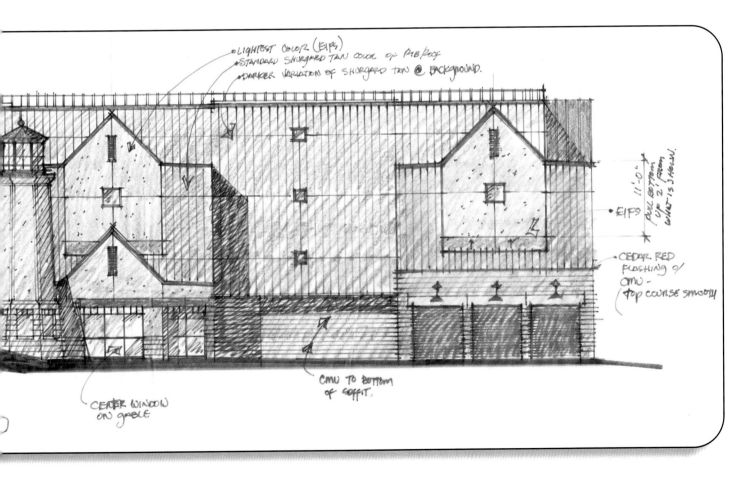

Design Tools for Engineering Teams

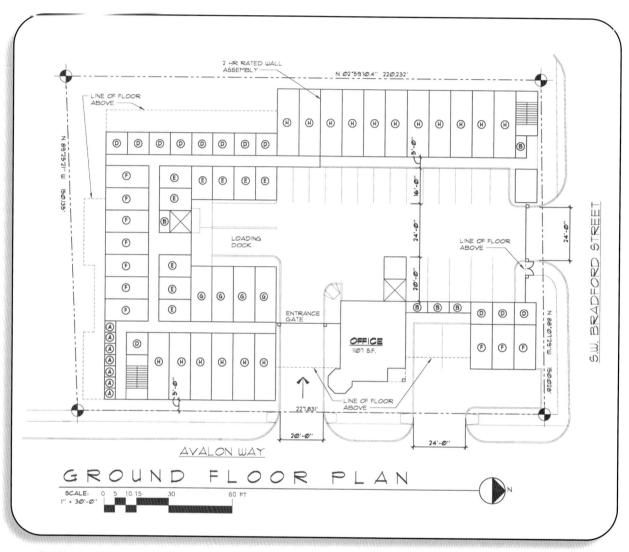

(© 2001 Shurgard Storage Centers, Inc. All rights reserved.)

Finished floor plan, ground floor

This print is of the finished floor plan for the ground floor of the facility. It shows a much larger parking area and revised parking configuration than that of the initial conceptual sketch.

(Drawings and images courtesy of Shurgard Storage Centers Inc.)

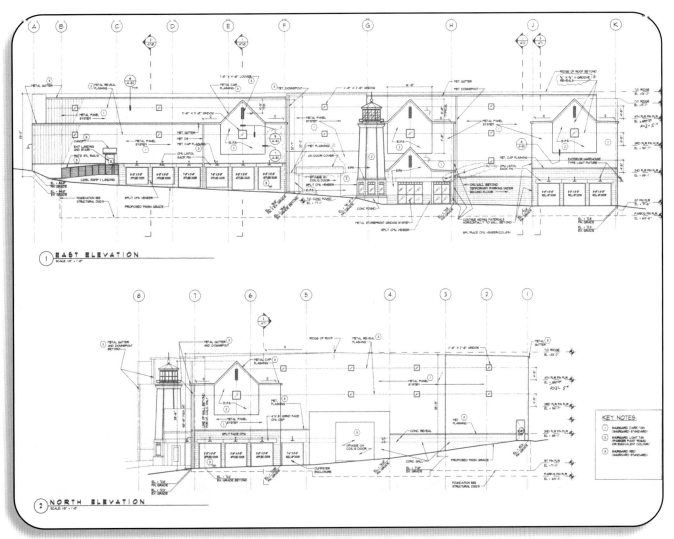

(© 2001 Shurgard Storage Centers, Inc. All rights reserved.)

Finished east elevation

This page shows the finished east elevation of the facility as it is to be built.

(Drawings and images courtesy of Shurgard Storage Centers Inc.)

Design Tools for Engineering Teams

(© 2001 Shurgard Storage Centers, Inc. All rights reserved.)

Completed facility

This and the following photograph show the the east elevation of the completed facility from different perspectives. Note the final configurations in the previous architectural renderings and how they look as a finished product. This project took approximately two years from the initial concept sketches to the finished construction. As it evolved and changed due to costs, aesthetics, zoning laws, company standards, and management revisions, many people from many parts of the organization contributed to its final design.

(Drawings and images courtesy of Shurgard Storage Centers Inc.)

(© 2001 Shurgard Storage Centers, Inc. All rights reserved.)

Completed facility

In the end, this particular project received many awards and recognition from the city of Seattle and the self-storage industry (1998 Facility of the Year) for the company's efforts to build a well-constructed, attractive state-of-the-art self-storage facility in a neglected part of town. Only by addressing the needs of the community, the business, and the customer, was the project able to succeed.

(Drawings and images courtesy of Shurgard Storage Centers Inc.)

Designing the Ford Focus

At Ford Motor Company a new vehicle design follows a set pattern from design concept sketches to final design and production. In the following illustrations let's track the development of the Ford Focus through the different stages of design and tests to a final car that is ready for the public.

Designer's concept sketches

For new concepts or major design changes, the design process begins with the designer's concept sketches. Exterior concepts are developed on paper until several are selected to be modeled in clay. The first models are one-fifth scale. These models are continually reviewed and modified to improve the design. For some vehicle programs this process is done via electronic model.

(Courtesy Ford Motor Company)

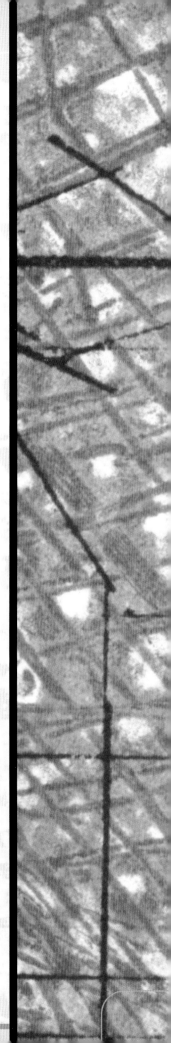

Design Tools for Engineering Teams

Full size clay model

From the initial clay model, concepts will be selected to be represented by full size clay models. These models will continue to be developed until the program is officially approved for production. At this point, a single model is selected and the final stages of the design process begins.

(Courtesy Ford Motor Company)

Interior clay model

Simultaneous with the exterior development, interior designers and clay modelers begin to develop proposals for the interior. These proposals will continue to be developed until the program is approved for production.

(Courtesy Ford Motor Company)

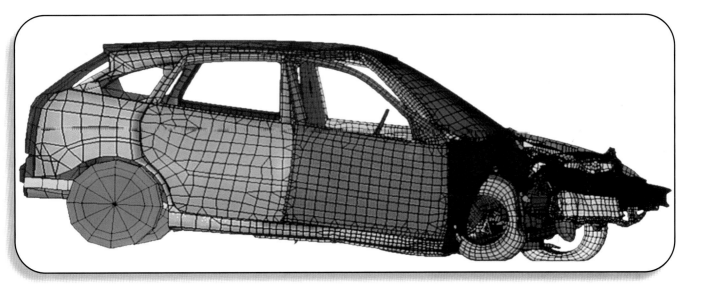

Computer crash prediction model

This is a computer model prediction of what would happen to the vehicle under certain crash conditions. These models allow the design engineers to evaluate a number of design proposals without having to wait for an actual barrier test. This significantly reduces the time required for this critical process.

(Courtesy Ford Motor Company)

Actual prototype crash test

This is the result of an actual physical prototype test and it is done at various stages of the product development process to validate the accuracy of the computer model. In this instance, there is very close correlation between the model and actual test.

(Courtesy Ford Motor Company)

Design Tools for Engineering Teams

Testing under actual conditions

Besides the many laboratory tests, vehicles must complete testing under actual conditions. Once the concept is finalized and design work is completed, actual prototype vehicles are built. This vehicle is being tested in Scandinavia for cold weather performance of the starting, heating, defrosting, and other systems.

(Courtesy Ford Motor Company)

Computer model of extreme lane change maneuver

This computer model is demonstrating an extreme lane change maneuver. By using computer-aided engineering models, engineers and designers are able to quickly review design alternatives and changes. This allows for more iterations of designs in a shorter period of time. Most of the design iterations are electronic so the cost and time to produce prototypes is substantially reduced.

(Courtesy Ford Motor Company)

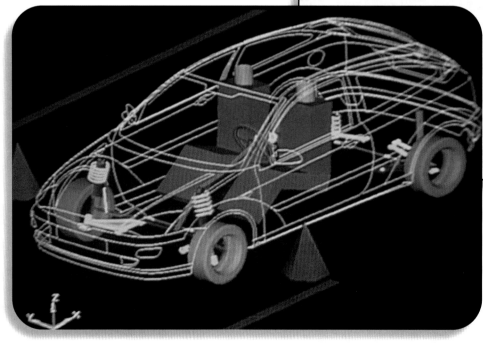

Design Tools for Engineering Teams

The finished product

With the design, development and testing complete, the car is ready for its public introduction.

(Courtesy Ford Motor Company)

A Town Built to the Customer's Specifications

In Frederick, Colorado, a once sleepy little front range town, the citizens responded to a 15.6 percent growth in population from 1990 to 1994 by requesting a plan for future growth and development of their fair town. The town of Frederick developed this comprehensive plan to guarantee this future growth and to meet the standards established by the citizens.

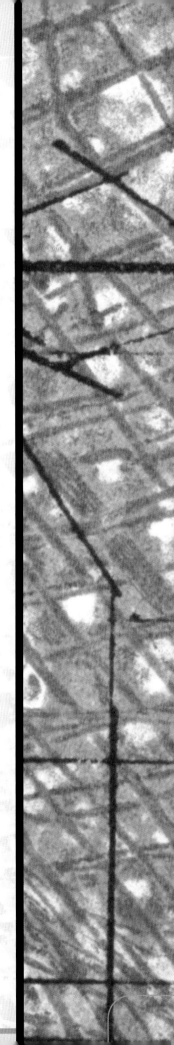

Contents of the Comprehensive Plan

The *Frederick Comprehensive Plan, 1996* has two components, a written and a graphic component. The written component addresses the following nine elements.

Growth of the Town addresses matters relating to growth including where, and what type of growth should occur in Frederick.

Image and Design of the Town deals with the quality of the physical environment and Frederick's "sense of place".

The *Housing* element of the plan covers supply of housing, housing diversity, and the quality of residential neighborhood environments.

Economic Development directs Frederick toward the establishment of a well-balanced economic base that provides jobs and economic opportunity for residents and a stable tax base for town services.

Transportation deals with street circulation, access, and alternative modes of transportation.

Public Improvements/Facilities addresses matters relating to cooperation with the Special Districts for the provision of public improvements and services ranging from utilities to libraries to fire stations; determining what is needed, the timing for improvements, and establishing funding priorities.

Parks, Open Space, and Recreation covers the types of parks and recreational opportunities needed in the Town and how to tie them all together as a system.

The *Environmental Issues* element of the plan identifies key environmental issues in Frederick and explores how they should be addressed.

Cultural, Historic, Educational, and Human Services Opportunities addresses the needs of Frederick residents in relation to community well being and quality of life.

Each of these elements contains a series of goals, policies, and strategies. A goal is the overall vision the citizens have for the community. A policy is a general approach the Town should take to ensure that the community is moving toward achieving the stated goal. And, a strategy is a specific method to implement the stated policy and ultimately, achieve the goal.

The graphic element of the plan is composed of the following series of maps:

- Land Use and Public Facilities Map
- Image and Design of the Town Map
- Transportation Map
- Utilities Map
- Parks, Open Space, and Recreation Map
- Environmental Issues Map

These maps are a graphical representation of the goals, policies, and strategies presented in the written element of *The Frederick Comprehensive Plan, 1996.*

Contents of Frederick, Colorado's Comprehensive Plan

The Frederick Comprehensive Plan is an example of a design process that has the customer totally involved in all phases of the design process.

A comprehensive plan is the culminating documents and maps that record all the elements of the Town's plan. This comprehensive plan contains three elements: goals, policies, and strategies.

(Courtesy Rocky Mountain Consultants, Longmont, CO)

> **GOAL 2:**
> Create a healthy balance between housing, employment opportunities, availability of goods and services, recreation and cultural possibilities within the community.
>
> **Policy 2.1:** Plan neighborhoods as readily identifiable communities bounded by natural features such as drainage basins and open space or human-made features, such as arterial streets and railroads. The neighborhood size should be relative to the scale of the existing core community which is approximately ¼ of a section (160 acres).
>
> > **Strategy 2.1.1:** Plan residential neighborhoods that are self-contained identifiable environments, centered around parks and community facilities, all within walking distance of the home. Include convenience services to serve a combination of neighborhoods.
> >
> > **Strategy 2.1.2:** Plan the location of school sites to serve a combination of neighborhoods. In conjunction with the St. Vrain Valley School District, identify the general location of multi-neighborhood schools on the Land Use Plan Map.
> >
> > **Strategy 2.1.3:** Commercial and industrial neighborhoods should be functional, identifiable areas that do not adversely impact adjacent residential neighborhoods or other uses. Require projects to use appropriate landscape buffering techniques and setbacks.
> >
> > **Strategy 2.1.4:** Minimize and discourage land uses that would detract from the function and viability of commercial and industrial neighborhoods.
>
> **Policy 2.2:** Evaluate development proposals in conjunction with the Land Use Plan to assure a healthy balance of land uses for all neighborhoods within the Frederick Planning Area.
>
>
>
> SEE STRATEGY 2.2.3 LOCATIONAL CRITERIA
>
> > **Strategy 2.2.1:** Use the Land Use Plan as a general guide to land use decision making. In addition, evaluate development proposals for their potential to attain the goals, policies and strategies outlined in the Comprehensive Plan.
> >
> > **Strategy 2.2.2:** Development proposals should adhere to the Image and Design goals, policies and strategies and the Town of Frederick Design Guidelines.

Example of a goal

The first element is goals, or in this plan, the overall vision the citizens have for their community. The citizens of Frederick, Colorado, are the clients or customers in the design development process. They have been involved very early in the process and have been included in every phase of the development of this plan. The policies are the guidelines for the Town to ensure that the community is achieving its stated goals. A sample of one of the town goals is shown in Goal #2: "Create a healthy balance between housing, employment opportunities, availability of goods and services, recreation and cultural possibilities within the community."

(Courtesy Rocky Mountain Consultants, Longmont, CO)

A strategy that will help accomplish Goal #2

The strategies are the specific methods used to implement the stated policies. Strategy 2.2.3: "Consider the following when siting residential, commercial and industrial land use. This information should be considered in the context of the Neighbor Planning Criteria as outlined. . . ." This strategy exemplifies a specific method to accomplish Goal #2.

(Courtesy Rocky Mountain Consultants, Longmont, CO)

Strategy 2.2.3: Consider the following when siting residential, commercial and industrial land use. This information should be considered in the context of the Neighborhood Planning Criteria as outlined in Strategy 4.1.1 and help to determine the location of land uses within new developments to create functional neighborhood plans.

RESIDENTIAL

1. Consider utility capacity when determining the density of residential development within neighborhoods. Areas with limited capacity should be developed at lower densities with development clustered to provide for more efficient delivery of services.

2. Physiographic constraints and environmental (i.e. potential subsidence and floodplains) should also help to determine the density and location of residential development within the neighborhood. The units should be clustered on the portions of the property that are not impacted or where impacts can be mitigated, leaving the remainder of the property open.

3. Residential development should generally be located within ½ mile range of shopping, schools, and public or private parks or open space. All residential developments should give careful consideration to safe pedestrian access to these facilities.

4. Consider access and proximity to local, collector and arterial streets when determining the location of the type of residential unit within the neighborhood. Areas that generate higher traffic volumes should have good access to collector and arterial streets without traveling through lower density areas.

5. Consider compatibility with adjacent land uses when determining the location of different types of residential units within the neighborhood.

COMMERCIAL

1. Neighborhood Commercial
   A. Approximately 2 to 5 acres
   B. A trade area of approximately 3/4 of a mile
   C. A site which has good access, preferably located along a collector or arterial street

2. Community Commercial
   A. Approximately 10 to 30 acre site
   B. A regional trade area which serves a population of 10,000 to 15,000
   C. A gross square footage range of 100,000 to 200,000 square feet
   D. A site which preferably has frontage on two arterial streets

INDUSTRIAL

1. Industrial Development
   A. Access to major highways through the City's arterial street system with minimal travel through other land uses
   B. Compatibility with nearby land uses
   C. Proximity to other industries
   D. Sites with no extraordinary constraints to development where existing or planned utility service is available

Conceptual drawing of Strategy 2.2.3

Through town meetings with the local citizens, the Town Planning Commission, the Board of Trustees, and Regional and State Planners, the goals were established and research on the town demographics was conducted using State and Regional data. A Civil Engineering firm did the research and development of the plan. Engineers and technicians were also part of this development team.

(Courtesy Rocky Mountain Consultants, Longmont, CO)

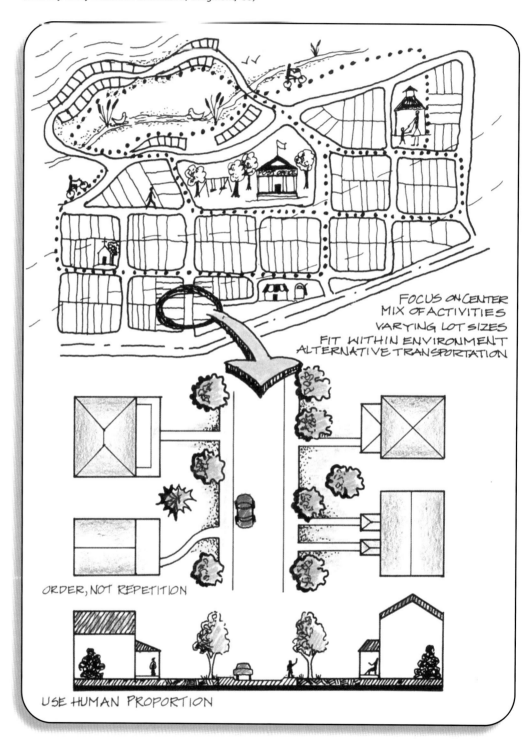

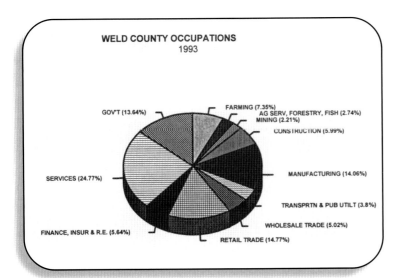

Sample of demographic data

Research on population trends, occupations, and earnings was conducted to have real data to build this plan.

(Courtesy Rocky Mountain Consultants, Longmont, CO)

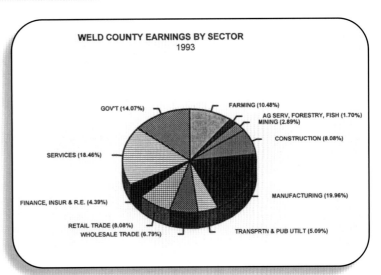

Population Trends							
	1980	% Change (1980 - 1990)	Avg. Annual % Change	1990	% Change (1990 - 1994)	1994	Avg. Annual % Change
State of Colorado	2,889,735	14%	1.4%	3,294,394	10.9%	3,655,647	2.7%
Weld County	123,438	6.8%	.7%	131,821	9.1%	143,824	2.3%
Town of Frederick	855	15.5%	1.6%	988	15.6%	1,142	3.9%
Source: Colorado Division of Local Governments; U.S. Bureau of the Population: 1980, 1990; Colorado State Demographer's Office							

Design Tools for Engineering Teams

Image and design of the comprehensive plan

In this design process the customer became the driving force behind the design decisions. The engineers and technicians were the support team to build a design based on real data and knowledge of land planning.

(Courtesy Rocky Mountain Consultants, Longmont, CO)

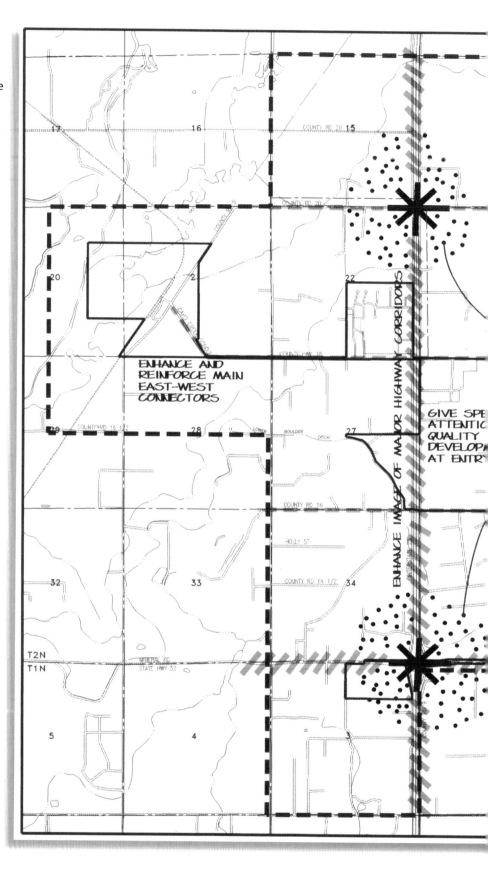

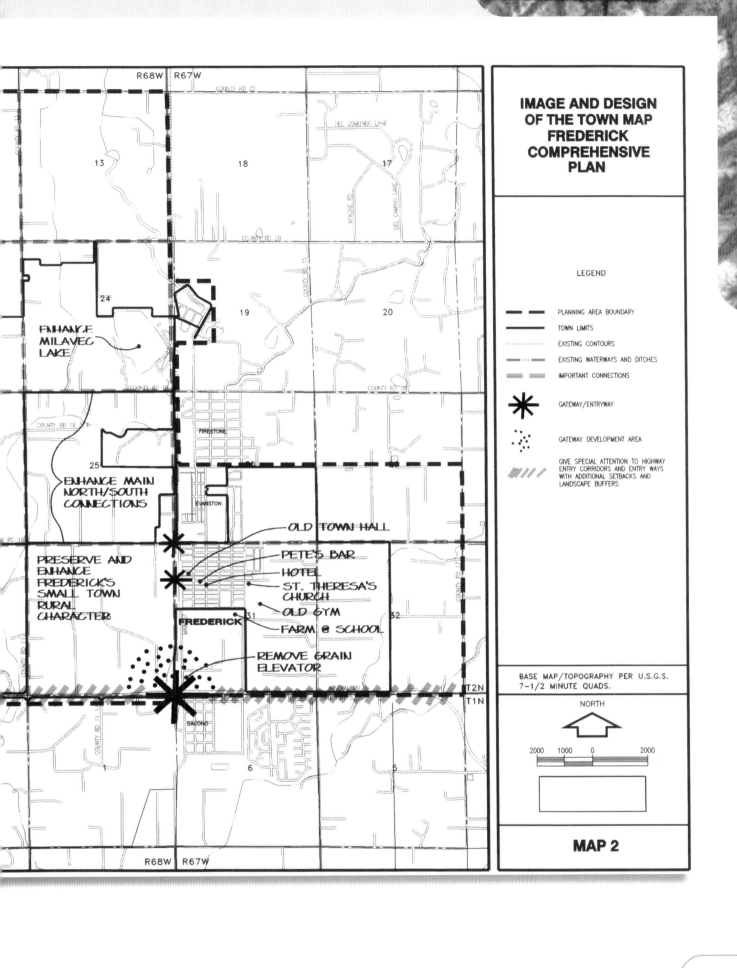

Design Tools for Engineering Teams

Land use and public facilities map of Frederick, Colorado

The map shown here reflects the collaborative results of the design process.

(Courtesy Rocky Mountain Consultants, Longmont, CO)

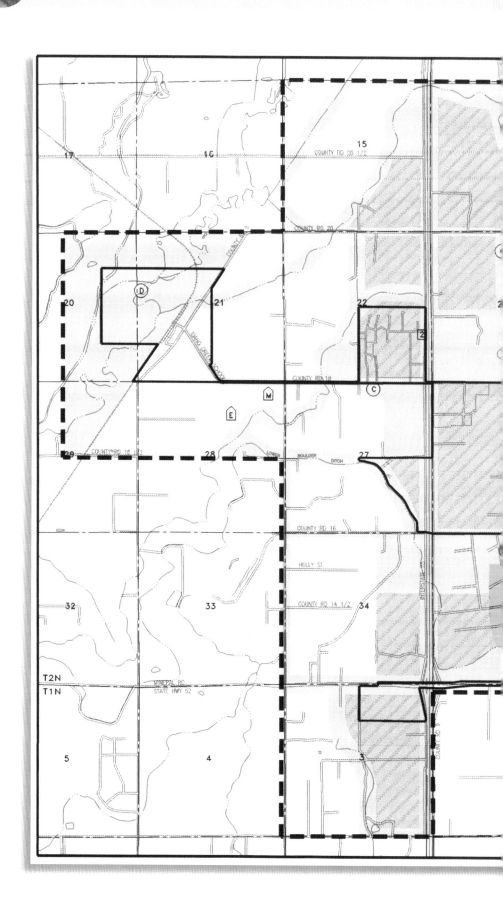

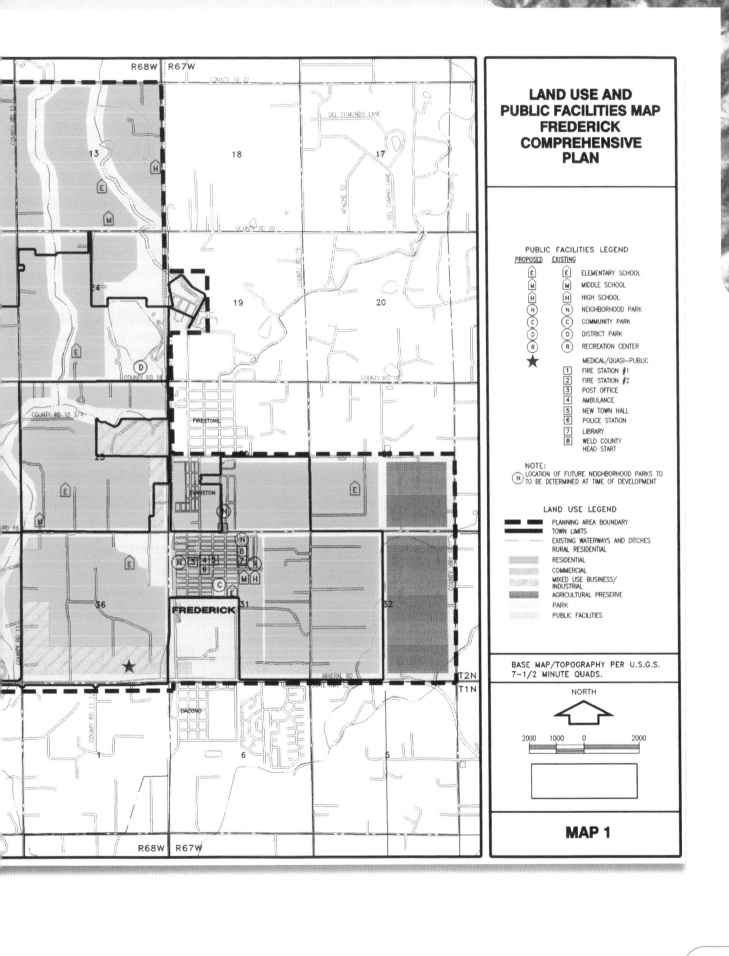

Design Tools for Engineering Teams

A sample form from the *Project Evaluation Workbook*

The results are policies to implement the Comprehensive Plan. In the case of Frederick, documents entitled *Town Official Project Evaluation Workbook* and *Town of Frederick Design Guidelines* were developed, as well as zoning and subdivision codes. Any new development project will have to submit a workbook and set of plans that meet the design guidelines to the Planning Commission and Board of Trustees for approval.

(Courtesy Rocky Mountain Consultants, Longmont, CO)

PROJECT EVALUATION WORKSHEET FOR TOWN OFFICIALS

Project Name _____   Type of Application _____

Applicant _____   Project Location _____

Name and Title of Reviewer _____   Date _____

SECTION ONE: PROJECT INTENT AND VISION

Issue for Review	Comments/Suggestions for Applicant	Project Strengths	Project Weaknesses
Do you have a good understanding of the intent and vision of the project? Key topics to look for include: A. The type of development B. The goals for the project C. The character of the project D. A reason why the applicant wants to develop in Frederick What are your thoughts about how this fits with what Frederick wants for the community?			

Problem-Solving Processes for Design

CHAPTER 4

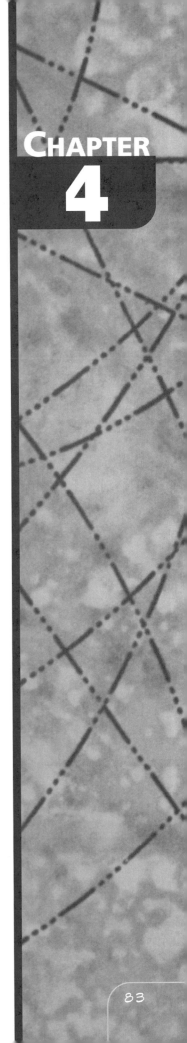

> "I keep six honest serving men (they taught me all I knew); their names are what and why and when and how and where and who."
>
> Rudyard Kipling

Introduction

Each and every day individuals are faced with problems or opportunities to choose a different way to do a job or reach a goal. A problem can be defined as a barrier that lies between us and where we want to go. A barrier creates a problem only when there is a goal.

In today's world of constant change, business reengineering, and the demand for "cheaper" and "faster," a person's ability to problem solve becomes a critical skill in the day-to-day processes of doing business. Hewlett Packard published a document about necessary skills of the workforce and it defines problem solving as "the ability of a person to have the basic skills to make wise use of available resources and information to be able to reason and make decisions in a perceptive and creative manner."

As the United States and international corporations are looking at how to survive in a global economy, skills of all workers are critical to the success or failure of these ventures. Corporations and workers require flexibility, agility, responsiveness, a strong customer orientation, the ability to explore and claim niche markets and create "built-in" quality. These are competitive requirements for the future that are built on foundation blocks of continuous innovation, collaborative workers, and creative problem solving.

The downsizing in the 1980s began the process of change in technical operations. This change produced a flatter and leaner organization that has removed layers of management and supervision. Responsibility for processing, analyzing, and decision making has moved to the workers at the lower level on the corporate ladder and the continuous rapid advancement in technology will further this phenomenon, since a person who processes and analyzes information will likely find in the future that computers are doing this job. The *interpretation* and *application* of this data will still be in the hands of the designer or engineer. Success with a company will be measured by the individual's skills to be flexible and creative, to solve problems within a system, and to acquire a broad knowledge of the total organization (see Figure 4.1).

Figure 4.1
Putting the puzzle together

Information Sharing

The increased availability of information will factor into the success of the individual to analyze data, draw conclusions, and present recommendations. Software will allow employees to make routine decisions that previously had to be taken to management. Being able to solve problems that arise in the technical operation or at the field site is essential for a firm's continuous improvement efforts to stay competitive.

The ability to solve these problems is enhanced by the employee having the corporation's shared information. This shared information on competitive strategies,

Chapter 4 | *Problem-Solving Processes for Design*

financial performance, corporate goals, and so on, has been traditionally kept secret for only a few at the top of the ladder. Now and in the future the workers will be expected to know their companies' goals, problems, successes, and objectives for the future. These same workers will be expected to act on this knowledge for their own personal performance, their team performance, and the performance of the company as a whole. These informed workers now have the knowledge to perform effective problem solving for the company, and this increased openness of information provides workers and worker groups with the ability to measure and track to see if their efforts are effective. As well, the free exchange of information brings management closer to workers by creating a greater sense of partnership. When the structure of hierarchy flattens, workers become associates.

Along with these benefits also comes an increased accountability and responsibility for the use of this information. The average designer can no longer create a design or prototype without consideration of market needs, costs or quality issues. The drafter, technician or designer can no longer blame engineering management or sales if the product fails. The responsibility is shared and along with that, the responsibility to contribute to the problem-solving process that moves barriers away to reach the corporate goal (see **Figure 4.2**).

Figure 4.2
Knowledge is shared

A structured and applied problem-solving approach will increase the success factor for the individual employees, functional work groups, and employee teams to solve problems efficiently and effectively. This chapter will cover the information you need to be successful in the business of solving problems.

Evolution of Problem Solving to Achieve a Quality Product

The question becomes, "Why problem solve?" The answer is to meet the need for continuous improvement and demand for quality. Problem-solving techniques have become an intricate part of the daily process of design and manufacturing.

The demand for quality has forced an evolution and focus on problem-solving techniques. Until the early 1980s, the engineering process used inspection and testing almost exclusively to detect defects and improve products. The emphasis was on "quality control" rather than "quality improvement." This approach died a slow death as a result of the high costs of the inspection process and cost of rework, scrap, and paying off warranty claims. Bad quality was inspected out, not designed and built in.

Industry Scenario

D&G Sciences Problem Solving: The Estimate Approach
D&G Sciences developed a novel problem-solving technique applicable to protracted problems.

The Premise: Certain problems can be solved faster, if an estimate of the effort needed to solve the problem occurs.

Most problem solvers skip the step of estimating the difficulty of the problem they are trying to solve; or if they do estimate it, they take that estimate lightly. The Estimate Approach claims that by investing in developing a good estimate of the effort to solve the problem, the overall solution effort will be reduced and perhaps minimized. In other words, the estimation will pay off by generating useful knowledge early and preventing futile sidetracking.

However difficult it may be to estimate the effort to solve a certain problem, it is an easier task than finding that solution itself.

The estimate approach is particularly powerful for long-term, large team research and development projects.

> Illustration: In 1989 an enterprising construction company developed a solution for preventing certain landmarks in Venice, Italy, from sinking further into the sea. They could not get funded because they could not produce a credible estimate of what it would take to execute their solution, which involved some new concepts of injecting a combination of new ceramics and plastic bonding amalgam into the old wooden foundation. Eventually the company addressed itself to a simpler problem; how to do the same if there was no water around (only the sinking piles). A pilot case was constructed, the injection process was credibly estimated, and then allowance was given for the water factor to arrive at a good total estimate. The need to produce a credible "dry" estimate led to significant improvement in the solution methodology. Had it not happened, the original plan would have stalled, run a huge bill and probably terminated prematurely.

In the early 1980s, the move shifted from test and inspection to statistical process control (SPC). This approach focused on training operators to monitor quality through control charts to determine trends in process performance. Variations in the product specifications were charted and samples checked. Problems were then turned over to an employee problem-solving team. Brainstorming and cause and effect charts were some of the tools used to identify potential problems and the cause. Action plans were developed to correct the problem. One downfall of this process is that the problem was barely analyzed as to cause before jumping into brainstormed solutions. The process was not systematic to effect consistent quality changes.

At this time while the United States was adopting SPC, the Japanese were moving from SPC to a new approach called design of experiments (DOE). Design of experiments (DOE) is a formal troubleshooting and problem-solving process that utilizes a controlled experiment to discover the important variables in a product or a design. This approach was to find specific causes of variations in product specifications as opposed to isolating out-of-tolerance situations as in SPC. SPC tolerates a certain amount of variation in product characteristics and DOE does not. The standard that has been traditionally accepted in design is to accept a variation in design specifications. For example, a bracket would have a tolerance factor on its length of 10 inches ± .005 inches. SPC accepts the product as long as it falls within that tolerance zone. Design of experiments (DOE) process expects the length of the bracket to be exactly 10 inches (see Figure 4.3). Under DOE, the objective is to eliminate all or as much variation as possible from the target value. The assumption is that customers want uniformity of quality in their products. Without a systematic approach to problem solving, Murphy's Law will prevail.

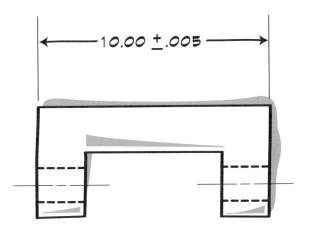

Figure 4.3
Design specifications

How does this impact problem-solving techniques? SPC used a traditional problem-solving process and basically used trial and error adjustments to improve the process. Since DOE is more systematic in its approach to problem solving and is highly technical, DOE helps technical operations move closer to producing products that nearly match their design in test batch after test batch. Therefore, DOE is generally a design and problem-solving tool. SPC is a process-monitoring tool.

Design Tools for Engineering Teams

One of the other discoveries of the 1980s was that "the customer is always right." The customer defined what quality was by accepting or rejecting the product. The expectations then shifted from the product meeting specifications, to meeting the customers' needs for quality. As a result, another quality improvement tool emerged: quality function deployment (QFD). Originating in Japan, a simplistic definition of this approach is first, decide what is important; second, design to reduce variation; and third, optimize the product. QFD has been highly held as the best process among businesses because it has reduced engineering changes by 30 to 50 percent, shortened design cycles by 30 to 50 percent, and significantly reduced warranty claims.

The premise of QFD is that the product's design matches the customer's requirements. This process is carried out by cross-functional teams that represent all aspects of product development from marketing, design and engineering, manufacturing, purchasing, and accounting. QFD taps into the collective knowledge of an organization and helps the design team catch costly problems and as many design flaws as possible as early as possible in the development process. This methodology helps teams make better decisions, and as a result, product development is much easier and less costly. The QFD process is tied to the customer's requirements, which are then translated to engineering and technical requirements, so QFD is primarily a product development tool.

> *Toyota and Honda release a new product every 3.5 years using the QFD methodology.*

QUALITY FUNCTION DEPLOYMENT

Use quality function deployment to:

1. Identify the customers vital requirements for the product or service
2. Develop a service blueprint of an elegant, effective, and efficient delivery process
3. Evaluate fault or failure points that need to be overcome in the design
4. Implement the newly designed process for delivery of the product or service

Quality function deployment consists of four main steps:

1. Identify the customer's vital requirements for the product or service and translate them into design requirements
2. Develop a service blueprint of an elegant, effective, and efficient delivery process
3. Evaluate alternative designs
4. Implement the newly designed process for delivery of the product or service

QFD is a planning process through communication of all aspects of the product development with documentation that seeks to guarantee the customer requirements through technical specifications. The utilization of quality function deployment (QFD) is a powerful problem-solving tool for reengineering a process. As processes become error prone and too slow to meet the needs of the customer, a business must incorporate new technology and processes to produce a better, faster, cheaper product.

Three different methods of improving a product or process have been reviewed here. One method looks at inspection, the second uses constant improvement of processes to improve the quality of the product, and the third is customer focused. All three of these processes utilize fundamental problem-solving tools and techniques. This next section will illustrate the many steps and tools used in a problem-solving process.

Industry Scenario

— James E. Folkestad, Ph.D., PMP, Assistant Professor, Colorado State University, Department of Manufacturing Technology and Construction Management

Each year students within the Department of Manufacturing Technology and Construction Management (MTCM) plan, design, manage, manufacture, and assemble toys for the Tots-for-Tots holiday toy giveaway. Students are taught to view this activity in a whole systems approach realizing that in order to get the job done they will need to work not just with classmates but with local community, university, student, and department leaders. Students must understand that failure to work effectively with anyone of these groups could delay or end the project in failure.

In order to help them cope with the demands of whole system management they are taught a variety of project management tools. These tools, many of which are described in the Project Management Institute's Body of Knowledge (PMBOK), provide students with the mechanisms to control the complexities of human interactions that take place during the engineering design process.

As an example, students are required to conduct a stakeholder analysis. This analysis focuses on identifying key project participants and laying out a process for communication. Students identify who needs what information and when they will need it. An example stakeholder analysis document follows.

Stakeholder Listing and Analysis

1. Project Manager and Dr. James Folkestad — Must be aware of the progress of entire project. Information needs consist of project progress reports and audits throughout the planning process.
2. Parents — Must be certain that product conforms with cultural and moral beliefs that the parents are trying to teach. Initial consultation will be done to narrow design specifications for product. Information needs consist of final product design(s) for verification and/or suggestions.
3. Even Start Family Center — Must be certain that product does not violate company standards or applicable laws. Initial consultation will be done to narrow design specifications for product. Information needs consist of final product design(s) for verification and/or suggestions.
4. Design Team — Must be aware of cost and time factors as well as product specifications and design constraints. Information needs consist of milestones for design(s) and manufacturing constraints.
5. Marketing Team — Must be aware of design features and project milestones. Information needs consist of consultation with parents, final product design with features, and dates needed for customer confirmation.
6. Manufacturing and Dr. Steve Schaeffer — Must verify that product can be manufactured using tools available. Information needs consist of product design(s), manufacturing process assumptions, and materials list.
7. Suppliers — Must be certain that suppliers will be able to meet the cost and time restrictions that production will require. Information needs consist of delivery dates, lot sizes, and payment process.

Design Tools for Engineering Teams

Industry Scenario, continued

8. Notebook coordinator — Must be certain that thorough documentation is given to the project manager and Dr. Folkestad on time. Information needs consist of all documentation required for notebook fulfillment.

After establishing this stakeholder analysis, one project team interviewed several parents of the children who received the toys. One of the parent's comments was related to a previously produced toy, a wooden duck. Her concern was that for the past three years at least half of her six children had received a wooden duck. Her worry, humorously expressed, was that she was running out of room for ducks. Again project management tools can be used to help teams facilitate whole system thinking (looking beyond their team) by working through a process and minimize unintended consequences.

Information that is gathered from stakeholders was then used to solidify design requirements and to maintain design intent throughout the project. Students develop a quality function deployment (QFD) chart that relates the "customer voice" to specific engineering design requirements (see the following example). Once completed this document is used to maintain project scope. Project managers can refer to the QFD to verify the importance of design features to the customer, the primary stakeholder. This tool helps project managers prevent scope creep (scope growth) inhibiting design team members from adding "nice to have" design features.

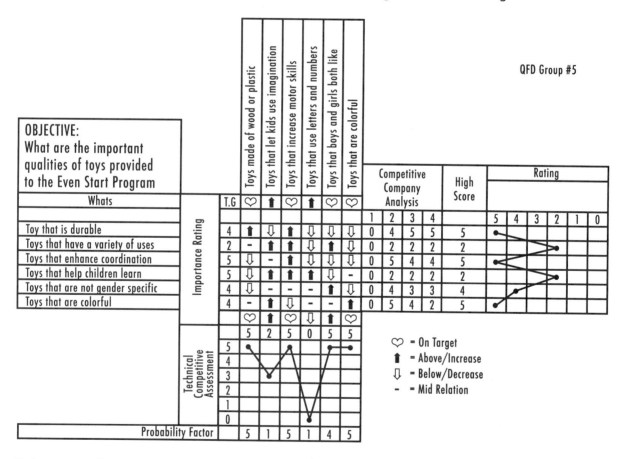

Each year Toys-for-Tots project managers document in a project notebook what they accomplished using their newly learned tools. These notebooks are developed using a standard format so that in subsequent years student project managers can read about previous toy projects and learn from successes and failures. Quality gains in both the engineering design process, including improvements in toy designs, have been witnessed over a number of years using these project management methods.

The Problem-Solving Process

Factors That Support or Hinder Problem Solving

Problem solving can be a lengthy and very complex process since there are many factors that impact the success or failure of any given problem-solving process. The success of this type of continuous improvement process is directly affected by your ability to identify the problem, analyze the data, and use effective tools to eliminate the problems that are encountered.

In many companies the inability of engineering or manufacturing to effectively identify the root of a problem and permanently do away with it is a major obstacle to their success in the market. Problems that keep coming back increase costs, decrease quality, and ultimately impact customer relations.

Problem solving in business can be hampered by the attitudes of employees and/or practices established by management. The following are just some examples that block the efforts for true problem solving and represent a "Band-Aid approach."

- Attacking the symptoms, not looking for the root of the problem
- Dictator type management — decisions by authority rather than a team approach
- Decisions by opinions rather than solutions based on data
- Blaming the next guy up or down the line
- Jumping to conclusions
- Not getting enough of the right information
- Attempting to solve problems that are beyond the team's control
- Solving one problem and creating another

These types of approaches to problems that arise create a "go nowhere" type of environment and, ultimately, failure.

Problem-Solving Steps

To be successful and overcome these natural tendencies, a company or team must share a common thought process for effectively finding solutions to problems. While there are many methods for solving problems, they all share four key principals[1]:

1. The problem must be clearly defined, so people know what problem is being solved and what the successful resolution of the problem will look like.
2. Merely "solving" symptoms must be avoided; everyone must focus on identifying and eliminating the underlying causes of the problems.
3. The chosen solutions must eliminate the problem and not cause additional problems in the future or in other places in the organization.

---

[1] Peter Capezio and Debra Morehouse, *Taking Aim on Leadership* (Kansas City, MO: National Press Publications, (800)258-7248).

4. Once fixed, problems must stay fixed. The organization must track and measure solutions to problems.

As we look at the components of a problem-solving process, let's look at another definition for problem solving and decision making.

> Problem solving "Demonstrates the ability to analyze and solve complex problems. Deals effectively with large amounts of data, changing conditions, incomplete data or uncertainty. Recognizes how seemingly unrelated issues or events interact and affect one another. Gets to the essence of complex problems quickly and generates effective courses of action. Is able to identify critical information from a wide range of data and use it to make effective decisions. Accurately interprets and understands quantitative and/or financial data."[2]

This definition prompts one to realize the different and interrelated parts of problem solving. It is a process. It is a process that can be defined. It is a process that can be followed systematically.

There are multiple ways to break down the steps to solving a problem. In a simplistic approach one must:

- **Identify** the problem
- **Define** the problem
- **Research** the cause
- **Explore** solutions
- **Act** on a solution
- **Observe** the solution
- **Evaluate** the success of the solution

This builds a systematic approach to linking problem solving and critical thinking. The first step is to identify the problem. The second step is defining the problem by working at clearly determining the specific information needs related to the identified problem. Detail the specific causes of the problem and work with local experts that know the surrounding conditions that could have contributed to the problem. It is very important to seek input from people at all levels. Let's take the example of designing a car. The levels of people to consider when problem solving a specific design idea would be the user (customer), the assembler, the designer of the assembly machines, the material vendors, marketing and accounting. One design change impacts many levels of the total product. One idea can change the assembly procedure, adjust the cost, and determine if it would be acceptable by the customer. When defining the problem, list all of the assumptions about the situation before jumping to conclusions. Bring the facts about the problem to respected experts for their viewpoint on the situation.

---

[2] Peter Capezio and Debra Morehouse, *Taking Aim on Leadership* (Kansas City, MO: National Press Publications, (800)258-7248).

Once the problem is articulated, attention turns to looking at a range of possible solutions. Utilize a group of people familiar with the situation to gain different perspectives. Brainstorming ideas develops alternative solutions and grouping ideas in chunks begins to show meaningful patterns. These patterns can lead to solutions. Mapping relationships within the information can also show patterns and points where action can be implemented. In looking at possible solutions, consider the perspective of other departments or work units. This avoids creating a solution that may create another problem elsewhere.

Creativity is a good friend of problem solving.

Generate several alternative courses of action then prioritize in terms of practicality, cost, and other factors that impact the success of the solution. Select an ideal solution to the problem and work backwards to develop the appropriate steps for implementation. Then look at the "worst-case scenario" for the solution that was chosen to implement. Design contingency plans based on this situation.

Put in action the selected solution and plan. Give it some time to work. It is easy to jump to conclusions at the first signs of failure. Sometimes it takes time to work out the kinks. After a respectable length of time, ask people who were involved with the definition process of the problem to give their critique of the potential solution. Identify specific strengths and weaknesses and determine any developing pattern such as frequency, data, voids, and so on. Based on these observations, develop a plan, chart, or spreadsheet to chart out the pattern. Methods for measuring the solution should be designed and agreed upon before the solution is implemented.

When evaluating the success or failure of the solution, weigh the pluses and minuses by outlining the consequences of the solution. Was the problem solved? Were the information needs met? Was the decision made? Was the situation resolved? Does the product satisfy the requirements as originally defined? The evaluation process determines how effectively and efficiently the problem-solving process was conducted.

With evaluation, the loop of problem solving could begin again. If in the evaluation additional problems are identified and defined, the problem-solving process begins all over again. It is a continuous process for improvement that is put in place when a predefined and agreed upon process for problems exists.

Industry Scenario

Decision Analysis

— *James E. Selle, Senior Instructor, University of Colorado, Denver Campus*

In all facets of life, decisions have to be made. Many decisions are automatic, such as those encountered while driving an automobile, and are based on experience. More complex decisions are also encountered such as which employment opportunity to accept, or which product (out of several options) to purchase. On the job, decisions such as which software program to buy for a particular application, or which person to hire for a particular position in your company. The ability to make good decisions is crucial to one's professional as well as personal well-being.

When more than one person is involved in decision making, the process becomes complicated. Without an accepted, unbiased procedure in place, decision making can become an exercise in which the person with the most power, the biggest mouth, or the strongest personality "wins." This results in poor decisions in which everyone shares the blame, not just the "winner." However, with a reasoned, sensible, accepted procedure in place, decision-making becomes a team effort in which everybody "wins." When properly applied, these procedures can ensure that differing positions can be resolved and the resulting decision can be unbiased. A number of decision-making procedures have been developed. The system described in this discussion is patterned after the one proposed by Kepner and Tregoe (see note 3).

The following thinking patterns should be used:

1. Appreciate the fact that a decision needs to be made.
2. Consider specific factors that *must* be satisfied if the choice is to succeed.
3. Decide what kind of action will satisfy these factors.
4. Consider the risks associated with the final choice.

A good decision can only be made in the context of "What is it that needs to be accomplished?"

First, a decision statement needs to be prepared. This statement provides the focus for everything that follows. It needs to indicate action along with its intended result ("Select a new Production Manager," or, "Devise a personnel evaluation system"). In addition, the decision statement should indicate the level at which the decision is to be made (those participating in the decision-making process).

The objectives of the decision are divided into two categories: MUSTS and WANTS. MUST objectives are mandatory; that is, to come to a successful decision, these objectives *must* be achieved. Any option that does not satisfy a MUST is automatically eliminated, without exception. MUST objectives must be measurable. For example, a MUST objective may be: "The item must cost less than X dollars." Any alternative costing more than this is automatically eliminated from further consideration.

WANTS are objectives that will be judged on their relative performance. A WANT may be mandatory, but may not be measurable, and cannot provide a yes-or-no answer. It is also possible that we do not want a yes-or-no answer, but a relative measure of performance.

Once alternatives have been identified (for example, candidates for a position, or an activity), information is essential. After all, it is folly to make decisions without adequate information. This information is then used to evaluate the various WANTS and to provide relative comparisons of the WANTS with each other.

After a choice has been made, the consequences of the choice need to be addressed. These consequences should address the *probability* of occurrence (high, medium, or low), and the seriousness of the consequence (high, medium, or low), should the possible consequence become a reality. The consequences could be "The chosen supplier may not be able to provide technical support," or "The company is a small company and may have trouble meeting out delivery requirements," and so forth. Obviously a high probability and high seriousness are to be avoided. These reviews are essential to a good decision.

Chapter 4 | Problem-Solving Processes for Design

Industry Scenario, continued

An example of a decision analysis is provided in the situation involving the purchase of a new automobile. The decision statement in this case is "Decide which automobile to purchase that meets my needs." The decision is to be made by the purchaser. The purchaser listed the following MUSTS:

1. MUST have front wheel drive
2. MUST cost less than Y dollars
3. MUST *not* be made by manufacturer Z (bad experience)

A list of WANTS was identified. The WANTS should be determined before any alternatives are considered to minimize bias in the decision. Using the MUSTS, a short list was generated by an overview of available models, discussions with friends on their experiences, and a general "feel" about various options (see chart below).

From this analysis it appears that model B is the best choice, considering the WANTS listed. Other WANTS and different importance ratings could produce a completely different result.

The consequences of this decision are not severe. It would appear that all of the manufacturers will remain in business for some time; no radical changes in any of the models are foreseen. The worst situation would seem to be inaccuracy in the reviewers' ratings. Depending on the nature of the erroneous information, the seriousness could vary from low to medium, with a low probability.

Considerable effort was expended in the gathering of information on the various models. The following information came from professional reviewers on the Internet, various consumer books and magazines, and sales brochures. Most of the reviewers provide a judgment rating on various characteristics and features of the automobiles being reviewed. These were used whenever possible in the following table. The points for each WANT is the product of the importance and the score as determined by the various reviewers. The model with the most points is then considered best for the wants listed.

A sample of items that cause variation in a design or product process:

- Poor quality batches of material
- Defective equipment
- Untrained operators
- Variation in methods
- Cycle times
- Process management
- Support systems
- Materials
- Training

---
3 C. H. Kepner and B. B. Tregoe, *The New Rational Manager* (Princeton, NJ: Princeton Research Press, 1981).

DECISION ANALYSIS FOR AUTOMOBILE

WANTS	Importance	Model A Score	Model A Points	Model B Score	Model B Points	Model C Score	Model C Points	Model D Score	Model D Points	Model E Score	Model E Points
Good gas mileage	60	4.0	240	7.0	420	4.0	240	6.0	360	4.0	240
Low wind noise	40	6.0	240	8.0	320	6.0	240	6.0	360	4.0	240
Low engine noise	50	9.0	450	8.0	400	8.0	400	6.0	300	4.0	200
Good steering	40	8.0	320	8.0	320	6.0	240	6.0	240	7.0	280
Low maintenance	70	5.0	350	7.0	490	8.0	560	7.0	490	8.0	560
Good safety record	60	6.7	400	7.0	420	8.5	510	8.0	480	9.0	540
Overall rating	70	7.5	520	7.5	520	7.0	490	7.3	511	6.6	430
Total			2520		2890		2680		2621		2450

Design Tools for Engineering Teams

Problem Solving Is a Process

1. Establish the scope
2. Develop the project plan
3. Select the key performance variables to benchmark
4. Identify potential participants
5. Measure your own performance
6. Measure performance of benchmarking participants
7. Compare current data
8. Identify best practices and enablers
9. Formulate your strategy
10. Implement the plan
11. Monitor results
12. Plan for problem solving

> Storyboarding is a visual brainstorming technique. All ideas are written down and organized in groups on a board.

Problem-Solving Strategies for Teams

Consider the ways of the ant

Ants are the most advanced social insects. The industrious ant accomplishes tasks that are much larger than themselves. They have been observed carting an entire chicken bone back to their colony from some unsuspecting picnickers. How do they do this given the smallness of their size? Ants work together to accomplish and problem solve amazing tasks. What can we learn from the ants? That working together in cooperation for a common goal brings about results. One ant could not have budged the chicken bone, but together, a team of ants were able to move it a great distance. This cooperation brought about support, improvement, mutual help, and ultimately trust within the group.

This is the bottom line for effective team problem solving. For problem solving to work in a team setting, a common process is essential for sorting through problems and building useful innovative solutions (see Figure 4.4). Let's look again at the key principles of problem solving and indicate where factors for team applications should be applied.

Chapter 4 | Problem-Solving Processes for Design

Figure 4.4
Problem solving as a team

Problem Solving As a Team

1. Identify the problem

   There should be joint agreement that a problem exists. A written problem statement will help the team stay focused on the issues at hand.

2. Define the problem

   Joint agreement on the root causes of the problem leads to gaining agreement on the procedures to follow for possible solutions.

3. Explore solutions

   Brainstorming combined with problem-solving tools is natural for teams. Large numbers of ideas leads to a greater number of possible solutions. Brainstorming builds on the strength of the team through the sharing of ideas. Here is a suggestion to follow for team brainstorming:

   Generate a lot of ideas; then have the team prioritize them.

   a. Number each item.
   b. Let each team member delete, combine, or add new ideas.
   c. Let each team member give positive justification for their favorite ideas.
   d. Vote on each item by show of hands. Cut items that don't show significant votes.
   e. Repeat the process if necessary.

4. Act on a solution

   The key to success is that all members on the team are committed to support the idea and contribute to the process of implementation.

5. Observe the solution

   Utilize the different skills of each team member to build a body of data to effectively evaluate the results. Communication among team members is critical in this part of the process.

6. Evaluate the success of the solution

    Based on team input and observations, the team must reach consensus on the success or failure of the solution.

Advantages and Disadvantages of Group Problem Solving

Problem solving in a team has great advantages. Collectively the team has more knowledge than one individual. Out of this corporate knowledge is created a synergy that allows for the group to build on what each one knows individually. When we look at the design process and the utilization of cross functional teams, we find more cooperation and ownership on a problem-solving decision than if it were made by one or two individuals. Another advantage of problem solving in a group is there are more people to gather the data and a variety of viewpoints. This helps in covering all bases as well as seeing the problem from different perspectives. The end result is a better decision or judgment on the matter at hand.

As we consider problem solving in a team we also must be cautious. There are disadvantages as well as advantages. Group interaction can cause people to conform even when they really do not agree or see it as the group does. Decisions may come slower. The problem could become committeed to death. Another potential problem with a group decision is it may be dominated by one strong individual.

Nominal Group Technique (NGT)

A team can work on creating norms to avoid this. Also the problem-solving technique called nominal group technique (NGT) can eliminate individual domination. NGT is a method of generating group ideas and decision making. The idea generation is done with silent brainstorming. Then there is an individual voting process. Once the final list of generated ideas is determined, each team member places the ideas in rank order. The ranking for all members of the team would be combined. The ideas with the highest score would become the team's first choice and so on down the list. The individual ranking of each team member is combined to arrive at the overall team decision. The NGT process takes group differences and treats them equally. Each member has to follow the same process, so the dominant member cannot overwhelm the silent member.

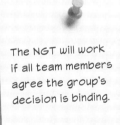

The NGT will work if all team members agree the group's decision is binding.

Another disadvantage of team problem solving is that a team can easily get off track. Without a good leader or facilitator, a group can stray from the original focus. Last, it is important to realize team members are human beings with personality differences. Generally when a group is problem solving, it is under pressure. This is a time when people can show their worst side. A team leader or facilitator can help the group during these stressful times to stay focused on the task and not on each other.

Chapter 4 | Problem-Solving Processes for Design

Industry Scenario

Community Teamwork

— Brad Breckenridge, Loveland, Colorado

Brad Breckenridge is a new civil technician at a civil engineering firm in a small town in Northern Colorado. His first assignment was to become familiar with the local procedures for developing a piece of property in their small town.

I gave a blank look when instructed by my supervisor to research municipal agencies and their roles with the land development process. I had no idea how that was done but I knew I would find out. Where do I start? How can I get information? Really now, how much could there be? A lot! After my first visit to the Mapping and Geological Information Systems department of our city, my blank look turned into awe. Nearly every agency used GIS. The visualization of information is a great tool that is now widely used. My next call went to zoning. There I asked, "How is your agency involved in new building and development?" After their long answer, I decided to sit in on one of the current planning meetings.

What I realized is the community uses a team approach to problem solving for land development. This team meets once a week in the City Building. Anyone can bring their project in to be reviewed with the team. Usually it is the landowner, architect, or designer that presents to the team. Any new development is first reviewed there. The setting is a conference room with visual media and sound recording equipment. The city provides representatives from the Planning, Power and Light, Parks, Emergency, Transportation, Water and Wastewater, Engineering, Natural Resources, and Historic Preservation departments. Each department has a name plaque and when called on, they give their review of the plan being presented. By the end of the meeting, the developer knows the errors in the plan. I observed that there usually were many errors. Questions were asked by the team such as Is the parking adequate and landscaped? What is your water runoff and wastewater plan? Are there proper easements for utilities? Should this light pole be moved?

My observations gave me the feeling that the process of property development is greatly improved when all the team players sit together at once to review a new development plan. Each agency then understands the requirements of the other agencies and everyone sees the whole picture. My first job assignment was a real eye-opener.

What Companies Are Doing to Facilitate Problem Solving

Some product development teams, which are comprised of all different aspects of the development process such as marketing, engineering, and manufacturing, are meeting with the whole development team away from outside interference, creating an environment that promotes focus and creativity. This freedom to work together promotes up-front problem solving and increases productivity.

Design Tools for Engineering Teams

Summary... Links to the Whole

Solving problems leads to a continuous improvement process. The more systematic the problem-solving approach, the better quality the solution. Problem solvers who generate and test more ideas are likely to develop better solutions. The production and quality of problem solving multiplies when implemented through teams that reflect the whole system of design and development of the product or project.

TOOLBOX

Tools to Use in a Problem-Solving Process

Is the problem continuous that involves systematic cause and effect across groups or organizations? Is the problem linear or does it have a direct effect on a product? Is the problem sporadic, infrequent, or a one-time problem? All these are questions and considerations as one seeks to identify the root causes that underlie a problem. There are tools and processes that one can select depending on the nature of the problem that can aid in reaching a solution. Listed here are some of the tools and their applications and an example of the tools used in a process. Selected tools will be determined by the type of problem to be solved.[4]

- **Surveys** — used to gather feedback from a variety of sources on one or more related topics

    Example: Survey the owners of a particular style of lawn mower to obtain information on gas usage, effectiveness of the cutting blade, ease of handling, and so on.

- **Data Printouts** — operational data of a particular system or machine (see Figure 4.5)

    Example: Read files in the computer that report error message in the software.

Figure 4.5
Data printouts

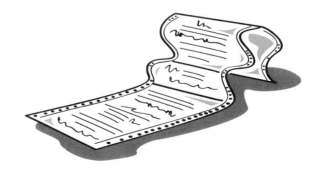

[4] (Denver, CO: Lifestar Publishing and Quantum Improvement Consulting, 2000), www.quantum-i.com.

Chapter 4 | Problem-Solving Processes for Design

- **Test Results** — test of performance or function (see Figure 4.6)

  Example: Pump test to see if the seal will hold the fluid going through the pump. Results could be the amount of leakage.

Figure 4.6
Test results

- **Observation** — actual visual observation of a process by a person (see Figure 4.7)

  Example: Person observes data output for a software analysis program.

Figure 4.7
Observation

- **Brainstorming** — a group individual process of thinking of all possible related ideas to the topic or problem at hand (see Figure 4.8)

  Example: Brainstorm all the possible ways a cell phone would fail to operate.

Figure 4.8
Brainstorming

Design Tools for Engineering Teams

- **Affinity Diagram** — identifies similarities between items (see Figure 4.9)

    Example: Create an affinity diagram to identify ways customer service training will benefit and affect customer service.

Figure 4.9
Affinity chart Diagram

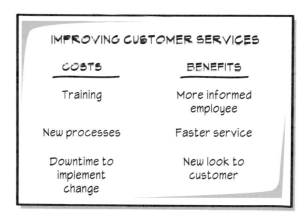

- **Nominal Group Technique (NGT)** — group idea generation and decision-making accomplished by ranking the choices through a silent brainstorming process (see Figure 4.10)

    Example: Use NGT to determine the best way to select an architectural style for a new subdivision.

Figure 4.10
Nominal Group Technique (NGT) (silent brainstorming)

- **Check Sheet** — data is recorded by classification (see Figure 4.11)

    Example: Use the check sheet to record all the safety settings on a machine.

Figure 4.11
Check Sheet

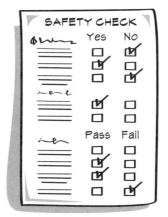

102

Chapter 4 | Problem-Solving Processes for Design

- **Histogram** — used to evaluate the capability of continuous data indicators (see Figure 4.12.)

  Example: One use of a histogram would be to show Fe_2O_3 concentration as related to rock quantity. The Fe_2O_3 concentration per rock quantity is charted. (Fe_2O_3 is a very common mineral, iron oxide, the principal ore of iron.)

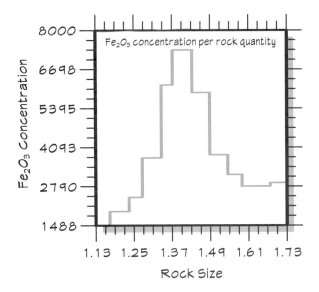

Figure 4.12
Histogram

- **Cause-effect Diagrams** (fishbone or Ishikawa diagram) — use the cause-effect diagram to identify the root causes of delay, waste, rework, or cost (see Figure 4.13.)

  Example: Use these to chart the cause of customer dissatisfaction as related to phone call-ins. Consider all the factors such as people, process, machines, and materials.

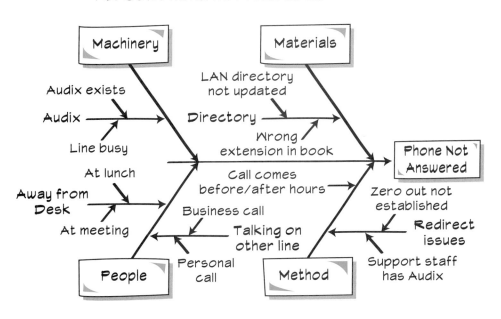

Figure 4.13
Cause-Effect Diagram

- **Pareto Chart (80/20 rule)** — a bar chart that is used to prioritize causes or issues (see Figure 4.14). The Pareto Chart combines a bar graph with a cumulative line graph. The bars are placed from left to right in descending

103

order. The cumulative line graph shows the percent contribution of all preceding bars. The Pareto Chart shows where effort can be focused for maximum benefit. It may take two or more Pareto Charts to focus the problem to a level that can be successfully analyzed. A common error is to stop at too high a level.

a. Identify the contributors to the problem: steps, machines, materials, or costs.

b. Sort the data in descending order (largest to smallest). Combine small quantities into others.

c. Draw the bar graph.

d. Draw the cumulative line graph on top of the bar graph showing the cumulative percent contribution.

e. Evaluate the Pareto pattern: Is the left-most bar significantly higher than the rest or is the pattern a Pareto? If there is a Pareto pattern, shade the bar to make it clearer.

Example: A Pareto Chart could be used to track the different causes of part rejection on a manufacturing production line.

Figure 4.14 Pareto Chart

REJECTED MANUFACTURED PARTS

defects = ['pits'; 'cracks'; 'holes'; 'dents'];
quantity = [5 3 19 25];
pareto (quantity, defects)

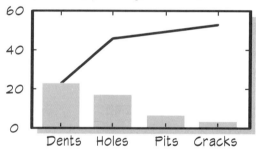

- Matrix Diagram — compares two or more groups of ideas, determines relationships among the elements, and makes decisions (see Figure 4.15). The matrix diagram helps prioritize tasks or issues in ways that aid decision making; identifies the connecting points between large groups of characteristics, functions, and tasks; or shows the ranking or priority of an interaction.

Example: Matrix diagram is like a simple spreadsheet. A construction company could plan out job priorities for each site.

Figure 4.15 Matrix

JOB PRIORITIES

Site Cleanup	Install Windows	Siding	Landscaping
Grounds 1st floor 2nd floor	1st floor 2nd floor Garage	Main house Garage	Front yard Back yard

Chapter 4 | Problem-Solving Processes for Design

- **System Diagram** — uses systems thinking to identify the systemic causes of delay, waste, rework and cost (see Figure 4.16). Selects cross-functional team members who have direct experience with the problem and its ongoing cost to the company, customers, and employees. Systems thinking recognizes that cause and effects are not always direct and linear (like the fishbone or cause-effect diagram might have us believe). Systems thinking is a way of looking at circular cause effects. In many ways, a systems diagram is similar to a relationship diagram.

  Systems Diagram symbols — boxes represent key indicators: the number, amount, or quantity of anything that can increase or decrease (e.g., amount of technicians, number of complaints, rewards, incentives, or time on a call)

  Example: The more pressure there is to "fix" service, the more resources we devote to fixing customer-affecting problems; the more resources we devote to fixing problems, the fewer resources we have to devote to preventative maintenance work, which ultimately leads to more customer problems and higher pressure to fix service.

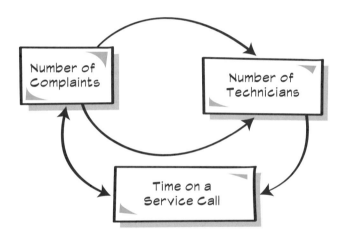

Figure 4.16
System Diagram

- **Benchmarking** — benchmarking for best practices (see Figure 4.17)

  The rigorous process of measurement and comparison of philosophies, policies, and practices against those in best-in-class organizations worldwide to achieve breakthrough improvement and surpass their performance. When financial benchmarking, ask What should our targets be? For strategic benchmarking, ask What strategy must I implement to achieve financial success? And for functional benchmarking, ask How does our operational performance support our strategy?

  Best practices are superior techniques that are dependent of industry, leadership, or management that leads to exceptional performance.

  Borrow from the best. Use benchmarking to determine "best-in-class" for a product or to perform a competitive analysis or initiate a benchmarking study to adapt the techniques of industry leaders in your business.

  Example: Benchmark a printed circuit board (PCB) design with a number of CAD software vendors to determine the best CAD solution for the PCB design.

Design Tools for Engineering Teams

Figure 4.17
Benchmarking

- **Action Plan** — identify who will do what and when (see Figure 4.18). Most improvements will require a careful plan to ensure they are implemented correctly and can be measured to evaluate their performance. Otherwise, it would be like scattering seeds over unturned soil and hoping for the best. A good implementation plan will identify:

  - What countermeasures to implement
  - How to implement them
  - Who will implement them
  - When they will be started and completed
  - How they will be measured

  Example: An action plan would be extremely helpful when planning the construction of a shopping mall.

Figure 4.18
Action Plan

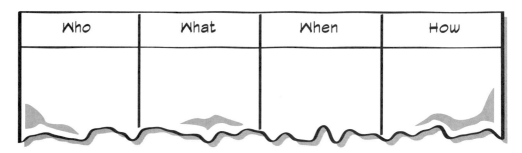

- **Action Chart** — used to evaluate the stability of the indicators (a machine performer)

  Example: This type of chart could be used to record data on a machine dimension. A company runs several shifts. They would record and measure 5 out of 100 parts to determine if there has been a drift in the tolerance of the dimension on any of the work shifts.

- **Force Field Analysis** — This technique is used to identify the forces that help (drive) and hinder (restrain) the implementation of a solution (see Figure 4.19).

106

Example: Use this type of analysis to outline the forces for and against upgrading a factory with new manufacturing equipment.

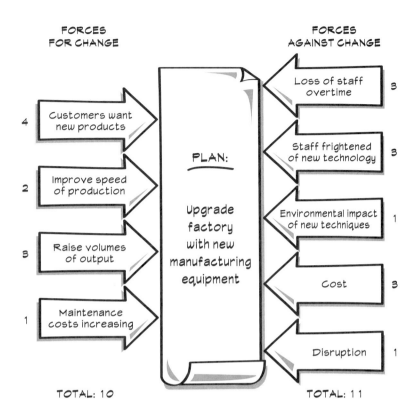

Figure 4.19
Force Field Analysis

When to Use Tools Available to Support the Problem-Solving Process

1. Identify the problem — The problem is brought forward from any party within the total system. The problem is identified as a condition or state that is different from the desired state.

   Tools: Survey Results

   Data Printout

   Test Results

   Observation

2. Define the problem — Once the problem is identified it requires a process to clarify it in terms of the local situation. The outcome of this stage is a well-defined problem statement.

   Tools: Brainstorming

   Affinity Diagram

   Nominal Group Technique

3. Research the cause of the problem — To research the cause of a problem it is necessary to explore the root cause. This is accomplished by collecting and analyzing data using statistical tools.

   Tools: Data gathering — surveys, interviews, reports

   Analyzing data — Check Sheet, Histogram

   Identify root causes — Cause and Effect Diagram, Pareto Chart

4. Explore solutions — Once the root cause of the problem is known then the team can explore the possible solutions to correct the problem.

    Tools: Brainstorming

                Nominal Group Technique

                Matrix Diagram

                System diagrams

                Benchmarking

5. Act on the solution — Once a solution is selected it must be tried and tested.

    Tools: Action Plan/Implementation Plan

                Control charts

                How-How Diagram

                Force Field Analysis

6. Observe the solution — Select best practices.

    Tool: Benchmarking

A-Team Scenario

The Pump Was Tested

Time has passed and the A-Team is well into the design of the pump. One design specification that continues to elude them is the design of a seal for the pump that will meet all the design criteria. This pump is to be designed to pump water or liquid pesticides. In addition, the pump has to withstand extreme hot and cold temperatures. The team realized that if the pump could pump liquid pesticides, it could pump water as well. So the focus went to the corrosive chemicals of the liquid pesticides. Usually the prototype pump would use water anyway to test the seals.

"Joe, you're our pump expert. What are our choices of seals?" asked Carlos, thinking this could be a simple question. "I'm not really sure right yet. I think we need to know more about the chemical composition of liquid pesticides," Joe responded. "We have many options when it comes to seals that will work in a pump. They range from the simple kind that is a graphic impregnated rope that is woven around the shaft and forced into the space to a seal that is a labyrinth type where two pieces work together to form an air barrier. There are also mechanical seals and lip seals." "Whew! I didn't realize we had so many choices, therefore so many decisions," replied Carlos. "You know, maybe we should bring this decision back to the group so we can have them help us problem solve this situation.

Joe quickly agreed. A group meeting was called and Carlos and Joe defined the problem they were working on at the moment. Everyone's consensus was that they needed to consult with someone who knew the chemical properties of liquid pesticides. Claudia said she would start working on that. Joe accepted the responsibility to check on seals and provide supporting data that explains how each seal responds to temperature variances and corrosive factors. Once all the data was in, Shantel placed all this information into a matrix diagram, which gave the team comparisons of the various seals. After the team worked with the matrix diagram, they were able to develop a cause and effect diagram for each type of seal. This really helped them make their decision.

A-Team Scenario
continued

"You know, I think it is time to do a trial run," Russell said as he sat back in his chair musing over the diagrams. "Let's send this puppy of a pump design over to the RPD guys and generate a part." "Yeah, right! Then we can place in the seal we have chosen and see if it really works," Carlos jumped in. "You know we need to be really systematic about this. Just popping a seal in and testing it isn't that simple," Joe warns. "Yeah, we know 'Design of Experiments!'" Russell exclaims. "My quality course has taught me how we should be doing this. I guess it is important to guarantee quality and consistency of the product." "We don't want to design a bad pump for Earl," Claudia responds. "Besides, I hate reworks. They just cost money and we don't have the time for reworks in this project."

They all got pretty quiet realizing how quickly their six-month deadline was moving in on them, but each felt good so far about what they had already accomplished together. Claudia set up the contract with the Rapid Product Development (RPO) firm nearby and Carlos sent his CAD model over to them by way of the Web. Joe ordered some of the seals they were going to try out based on the information they learned about the chemical properties of liquid pesticides. Shantel set up some spreadsheets to use in the design of experiments procedures. She made sure to accurately record all the various tolerances and standards for which these pumps and seals would be tested. Russell was beginning to copy off drawings from Carlos's solid modeling designs to develop the illustrations for presentation materials.

The day came when they were to test the seals. You could feel the excitement in the air. The A-Team invited Earl to come down to observe the tests. Earl welcomed the break from his ranch duties and was wondering where the project was at anyway.

"Hi, Earl. Welcome to our test session. How have you been?" greeted Claudia. "Doing fine and really want to see if the pump idea will really work," Earl replied. Everyone stood as close as they could to what was happening when they sent the water through the first pump. "Oh, no! The seal didn't hold!" Carlos blurted out. Everyone else was silent knowing their pump was being reviewed by Earl. Everyone just stood there for a few minutes, not sure what to say. Next, Joe started calling out tolerances, flow speeds, and volume figures related to leakages to Shantel. Earl had his face right down into the pump and he was listening to all Joe had to say. Then he walked around to the back of the pump and read the pressure gauge. Earl stood there for a few minutes and most of the team didn't even observe what he was doing. Next they hear Earl say, "Darn good design. This one is better than what I had." Carlos again said, "Yeah, but this one is leaking." Earl replied, "Well you'd leak too if you had this much pressure run through you." At this point they all focused in on the pressure gauge. Earl continued, "You have enough pressure here to pump water to put a fire out in a ten-story building. All I need is enough pressure to pump water into a ditch or pesticides on a field." Joe thumped his head and said, "You are right. We just didn't consider that variable. Let's set up the test again and adjust the pressure. Earl, you dial in the pressure. You are the expert on your applications." Earl did just that. This time the seal held. The team celebrated their success and recorded all the necessary data for future specification sheets in the final documentation.

After Earl had taken many photos of his new pump design, he left to go back home to share the news with his ranching buddies. The A-Team then met to go over their results and to figure out where they needed to go from here.

Carlos spoke up, "I guess Earl gives new meaning to the old adage 'The customer is always right.'" Claudia responded, "Earl knows the application and how the pump will be used. We just weren't thinking about that, were we?" "Now it makes sense what all the trade magazines are saying to involve the customer early in the design process. Just think of all the hours of testing and seal designs we would have gone through if Earl hadn't been here," Russell piped in. Shantel just cringed thinking of all that charting.

Design Tools for Engineering Teams

A-Team Scenario
continued

Respond to the A-Team scenario by answering the following questions. Use any tools or ideas you have learned in this chapter.

1. Identify the different steps of the problem-solving process that the A-Team used:

   a. Identify the problem
   _____
   _____

   b. Define the problem
   _____
   _____

   c. Explore the solutions
   _____
   _____

   d. Act on a solution
   _____
   _____

   e. Observe the solution
   _____
   _____

   f. Evaluate the success of the solution
   _____
   _____

2. What problem-solving tools were used in determining what type of seal to use?
   _____
   _____
   _____
   _____

3. What key element took place in the A-Team problem-solving process that helped keep time loss and costs to a minimum?
   _____
   _____
   _____
   _____

Chapter 4 | Problem-Solving Processes for Design

TERMINOLOGY & DEFINITIONS

Murphy's Law – common term that refers to the old saying, "What can go wrong will!"

out-of-tolerance (OT) – condition when a measurement on the product or process does not meet the specification

variable – any characteristic of a process or a product that may change

Activities and Projects

- **Identify** a problem that has occurred within your project and use the following steps to solve the problem. Document all steps and discuss them with your team.

    Determine the issues.
    _____
    _____
    _____

    Analyze the problem.
    _____
    _____
    _____

    Set a solution objective.
    _____
    _____
    _____

    Identify the criteria for a successful solution (all team members must agree to this criteria).
    _____
    _____
    _____

    Select the best solution. Give reasons why it is the best solution.
    _____
    _____
    _____

Design Tools for Engineering Teams

Activities and Projects *continued*

Plan how to implement this solution.

_____
_____

Review the results. Is the problem solved?

_____
_____
_____

- Write a reflection paper about the problem-solving process you used for the project problem or another problem you recently solved. In your reflections, consider how the team worked together, your areas of success, and areas that need improvement. Reflect on your own personal strategies for solving the problem.

_____
_____
_____
_____
_____
_____
_____
_____
_____
_____
_____
_____

- Determine which of the tools listed for problem solving will be incorporated into the team project design.

Problem-Solving Tool	What You Will Study
_____	_____
_____	_____
_____	_____
_____	_____
_____	_____

Communicating a Design: Making Your Case

CHAPTER 5

> "Be humble and thankful for the opportunity to express yourself."
>
> Anonymous

Introduction

Communication is the ability to convey thoughts and ideas verbally and with the written word, and to listen for understanding. With the changes of engineering and manufacturing business methods, comes an increased emphasis on the value of good communication skills. And with the fundamental change in the development and design of a product comes a basic change in how constituent groups work together and communicate.

Look at the definition of concurrent engineering: "A faster type of design engineering where the design, the manufacturing process, the materials, and the manufacturing tools are worked on at the same time to reduce manufacturing cycles and increase the likelihood of trouble-free manufacturing." This definition describes very diverse groups working together for a common goal, that goal being "improved product development."

Given what was discussed earlier about the involvement of the customer in the design and manufacturing process, it could be said that the customer requirements could be added to this definition of concurrent engineering. The customer needs to be part of the process. Without good communication between all parties in the design process, we could have a potential communication meltdown. It could be like walking into a restaurant where everyone is yelling out food orders at once without a wait staff to communicate the order to the cook. The results could be a meal that was not made to the customer's request because the cook never heard the requirements.

It used to be that a product evolved in a linear process. It used to be that each department operated independently with little communication except to tell the next group they were passing the project along. It used to be the orders and design rules came from top management. Today the rules of communication are changing. The design and development process requires a dynamic working relationship between all groups involved from the customer on up the line or, better said, across the system.

It is known that for concurrent engineering or system design to work, group interaction and communication occurs in the form of teams. The success of these teams will be measured in their ability to communicate. Team members will need skills to express their ideas effectively and share information accurately and quickly, as well as the ability to hear needs or problems. Good listening skills are as important as the verbal side of communication. It is like a seesaw (Figure 5.1), each being equal and requiring dynamic communication to work.

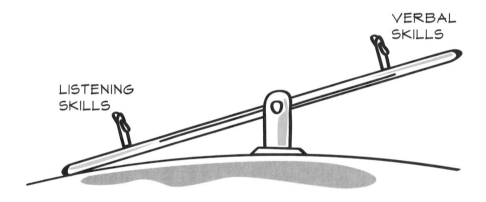

Figure 5.1
Good communication is a balancing act

When people work together their collective energy and ideas produce a powerful force, but it is only as strong as the communication link that connects them. Good communication gets things done. The success or failure of a product or project design is a direct correlation to how effectively the design team communicates with each other, across the system and with the customer.

Communication as a Strategic Planning Tool

For any group or organization to move forward with an idea or dream for an actual product, a plan for action must be in place. This strategy would include mutually agreed upon goals, standards for performance, and methods and expectations for communication. Communication, if not planned, can be random and haphazard which is generally crisis driven. This type of communication model breeds dissatisfaction, gossip and rumors, and a general demoralization of the group as a whole.

A successful organizational model will utilize communication tools to keep all parties informed in a timely manner, since the time spent in communication up front reduces the repair time for damage done when communication links are poor. Tools for communication will vary depending on the company size, project requirements, number of persons involved, and distance between groups. A small firm with two or three design teams will have very different needs for communication than a company the size of Ford Motor Company or IBM where design teams may be virtual and located in different cities or countries. The important point is that each level has a communication requirement and the tools chosen will reflect the best way to communicate. There are many effective tools for corporate or group communication. Some found in the communication toolbox (see Figure 5.2) are good examples of tools that foster communication.

Figure 5.2
Communication toolbox

A well-designed, thought-out communication strategy will become the oil that greases the wheels of innovation, problem solving, and realization of a vision.

Communication Is Essential to Success, but Why Doesn't It Work?

When we look at the diversity of the workplace and the constantly changing requirements placed on individuals, it is somewhat amazing that we have positive communication at all. The important word here is "positive." There is always a lot of communication occurring in a work environment, but it is not always focused or positive (see Figure 5.3).

Figure 5.3 Around the coffee pot: sometimes the wrong communication lets people down

To understand the necessity of using communication tools and procedures, it is important to recognize the factors that influence poor communication. General George Patton said, "Never tell people how to do things. Tell them what to do and they will surprise you with their ingenuity." What was General Patton trying to tell us? One, a person who is valued for their knowledge and skills can achieve great things. Secondly, when communication is only given one way from the top, it becomes a reactive tool. For true communication, both sides must be involved. When only one side is talking, the other side may or may not be listening. Do you remember a time a teacher was scolding your class about a situation that occurred? Chances are you were thinking of how to get out of there fast or how ugly the teacher looked when he was angry. Your reaction may have been different if the teacher had queried the group about why the class chose their response to a situation or what were the consequences of their actions. The questioning process involved the group and demonstrated respect for the group's decision for the action. The communication tool described is a questioning technique, which involves all parties in the discussion.

The right questions and the way they are posed create discussion. Dialogue occurs when questions are open-ended. Open-ended questions foster positive communication because they don't require yes or no answers. In addition, open-ended questions demonstrate interest on behalf of the person asking the questions. The use of open-ended questions allows the participants to express themselves with dialogue. Therefore, an example of an open-ended question would be "Where are you on the Global Solutions Project?" whereas a closed-ended question would be "Have you finished the Global Solutions Project?"

And so it becomes apparent that the better we know our team members or understand how the people in other departments think, the better the communication between each person or group will become. Today in the global market and with the emphasis on simultaneous engineering, people are working in a very diverse culture. Differences among team members could be in culture and background, age, gender, ethnicity, or work experience. To effectively communicate, we must focus on the task at hand and place value on the individual differences rather than perceiving these differences as negative (see **Figure 5.4**).

Figure 5.4
Communication is valuing the differences

Communicating positively and effectively will mean determining how the other person thinks, communicates, and operates. This communication tool is effective interpersonal relations. For effective interpersonal communication, one must first recognize the unique characteristics of others and support them. Why? Because people can cooperate in performing a task only to the extent that each performs work that complements the work of *others*. The breakdown occurs when focus is placed negatively on the differences rather than seeing the differences as positive contributions to the communication process.

When we look at systems thinking and linking varied parts of the system to achieve a goal, it is important to consider how to keep diverse groups of people connected. The glue that works is the goal, vision, or mission. This goal must be successfully communicated and embraced by all, and in the pursuit of the goal, individual differences are put aside or utilized as strengths to achieve the goal. As individuals work in different parts of the system (see **Figure 5.5**), the unifying factor is the goal that everyone is working together to achieve.

Figure 5.5
The goal keeps everyone together

Thus, communication is the tool to keep all parts of the system on task to achieve the goal. When the goal or vision is not shared or evident, communication across the system becomes fragmented and breaks down.

Tools that Promote Positive Communication

Positive communication will not occur when people do not feel safe to come forward with their ideas or if their contributions are not recognized. A "why bother" attitude begins to permeate the environment. It is like dry rot that slowly eats away at the system. People must feel valued for what they know and be encouraged to contribute their ideas and solutions. To do this, some of the communication tools that have been proven effective are quality circles, suggestion boxes, and employee incentive award programs, all emerging in the 1980s. These have led to communication tools for true employee empowerment through self-directed teams and more open communication with management.

Often we see suggestion boxes in places of business, doctor's offices, schools or marketplaces and we have wondered if the suggestions placed in these boxes were ever considered. Today, design and manufacturing teams are capitalizing on this idea of a suggestion box.

If a team member comes up with a new idea or a better way to do a job, it is submitted as a suggestion. If the idea is adopted, the person who made the submittal is presented with an incentive award. The incentive award provides encouragement to each participant to continue to come forth with creative new ideas. Incentive awards could be in the form of monetary awards or as simple as a certificate.

Quality circles have been successfully utilized in Japan, especially by Nissan and Toyota. They are formalized meetings of various members of a system. By working together and pooling their knowledge of a process, members of a circle are able to solve problems with a stronger base of information. In quality circles all levels from management to floor workers have equal power to contribute and all must work together to accomplish the same task. Communication works when everyone's ideas are heard and respected, and bringing management into the team as an active partner in the process is key to the success of communicating a design.

Making Team Communications Work

We have discovered that teams and team communications strongly influence systems design and concurrent engineering. Teams find themselves in an interesting position of communicating outwardly to the whole system as a unit and among themselves as a collection of individuals. Here we want to examine the inner dynamics of teams for effective communications.

For teams to effectively communicate, the overall purpose of their existence must be clear. This involves the understanding of what the team is expected to do and the goal they are to achieve. One way to focus this for the team is to have a team charter. A charter can describe how the team fits within and communicates to the whole organization. This charter helps each individual team member visualize his or her role in the team. In addition, it helps to set a basis for the team to make decisions and set short-term goals. The charter is developed by hearing every team member's voice (see Figure 5.6).

Figure 5.6
The team charter says it all

Another important factor for effective team communications is establishing norms. In Chapter 2, Teams as a Tool in the Engineering Design Process, we discussed the development of team norms and their importance. In this chapter we want to emphasize the norms that foster open communication among the team members. Some examples of communications norms are:

1. Each team member shares an opinion whether it is considered right, wrong, or indifferent.
2. No question is a dumb question.
3. All vital communications will be in written form.
4. Weekly or biweekly meetings are scheduled to consider any existing tensions.
5. Always verify total team understanding.
6. Celebrate and give positive feedback for good ideas.
7. Reach important decisions through consensus.

The way to verify team understanding is to ask questions, and team members should be encouraged to ask questions at anytime. Another tool for team understanding is progress reports. This tool lets everyone know where they are in achieving the goal. Sometimes scenarios can be used for communication. Scenarios are a great way to point out the "what ifs and whys" of a process or design outcome. A scenario could be used to help a design group identify possible problems a part or assembly could encounter under varying conditions. For example, a NASA space design group may create a scenario that simulates all the possible conditions a space station hatch may have to endure such as vibration on take-off, heat and cold variances, repairs in a weightless environment, and so on. Scenarios help flush out potential problems and are best used in problem-solving activities.

Reaching Consensus

In team communication, it is encouraged to hear everyone's opinions and allow all to have a voice. These opinions are very valuable, so that all have ownership in the process. For all members of the team to totally agree all of the time is probably not going to happen. The process recommended for a group to move forward on an issue is called *reaching consensus*. A consensus decision is one that the entire team can support although they personally may not agree with it. This type of decision is based on a common goal, upon which each team member originally agrees.

You can reach consensus by following these ten steps:

1. Define the purpose of the decision to be made by the team.
2. Use data and information to make intelligent decisions.
3. Team members come prepared with their own feelings on the issue at hand.
4. Team members share their thoughts about the issue.
5. Listen to all team members' viewpoints.

6. Clarify any unclear issues with questions.
7. Value differing opinions, which help to look at the issue from a different point of view.
8. Confront ideas, not people.
9. Work for a quality decision that all on the team can support.
10. Fully support the team discussion and individually take responsibility to see it through.

There are situations when consensus will not work or is not appropriate such as situations when the team is directed by higher management or they are working in a temporary crisis. Once the temporary crisis passes, the team should go back to doing a consensus process for reaching a decision. A consensus decision is the best for promoting teamwork and team focus. If consensus absolutely won't work, resort to voting and majority rule. The bottom line is to preserve team communication and morale.

Using Feedback for Furthering Open Communications

Teams, by nature, bring varying degrees of skills, abilities, and talents to the group. For a team to function as a unit, all must be open to receiving feedback as well as willing to give feedback to each other. Personal growth and team growth occurs when positive constructive feedback takes place. In a team, feedback is a two-way process that includes your point of view and the point of view of others (see Figure 5.7).

Figure 5.7
Team feedback

GIVING AND RECEIVING FEEDBACK

GIVING FEEDBACK	RECEIVING FEEDBACK
1. Focus feedback to team results.	1. Listen carefully to the major areas being addressed.
2. Base your perceptions on specific incidents and facts. Avoid opinions.	2. View the feedback as an opportunity for growth.
3. To avoid defensive positioning, involve the other person in the conversation.	3. Ask for clarification by asking questions and paraphrasing key points.
4. Develop a plan of action that is jointly agreed upon.	4. Prove your point of view of the situation while remaining objective and avoid making excuses.
5. Summarize the communication and the plan of action.	5. Offer your ideas for how to improve or change the situation.
6. Let other people know you appreciate their participation and openness.	6. Schedule a follow-up meeting to give a progress report.
	7. Thank the person for wanting to help you by giving you feedback.

It is important to remember that feedback can be positive or negative. How one gives feedback and receives feedback is critical to the trust and emotional health of the team. There are very few people who like to hear their performance is lacking and there are also people who do not like being singled out for a good performance! So feedback must be constructive and positive. The manner in which feedback is provided determines on whether the individual will use it to improve or rebel against the group.

The most important factor in giving and receiving feedback in a professional work environment is to keep it focused on the job at hand and not on personalities. A team is formed to do a job, not necessarily to be friends. It also helps to first practice giving and receiving feedback on easy items before the big crunch comes along. The team's skills will then be practiced and the sensitivity levels will be lower. The basic rule is to focus the feedback on behaviors or actions, not on people. Keep it short so it does not sound like a lecture and be sure to only speak for yourself and the impact on you or your job.

When receiving feedback, keep in mind the idea of mutual respect. Other people may see things from a different point of view and deserve the time to share their feelings. Out of this kind of dialogue of giving and receiving feedback can come some fruitful ideas that could benefit the overall project.

Running an Effective Meeting Can Improve Communications

Part of team communication is having group meetings. Routine meetings work to assess progress and maintain continuity of the group, however, the disadvantage of routine meetings is they can occur without a purpose or structure. Before the meeting, members should prepare an agenda, an outline that details the objectives or decisions to be reached, as well as the location, date, and start/finish items of the meeting.

The routine meeting can also become a place for members' personal agendas, so it is important to observe some basic rules for holding a meeting. *When* is it good for all to meet, *why* should the team meet, and *where* is it convenient for all to meet? Each meeting should focus on issues that affect the whole team and should be issues on which the team can have a significant impact.

Most people cringe when they hear there is going to be a meeting, but meetings can be effective if conducted in a way that focuses on the group. All members must feel their contributions are valued and that there is a sense of order to the meeting. These are items that can be determined early in the formation of the group. If the group is unfamiliar with how to use Robert's Rules of Order for parliamentary procedures then training is necessary. Small teams may not need such formal rules, but some guidelines or norms would surely help. Norms can be also written to facilitate methods of communication. For example, a norm could be that every member has 3 minutes to give his or her perspective of a problem brought to the table. This type of norm limits the time of input but fosters total group participation. Where a meeting is held and the seating arrangements also contribute to the success of a meeting. A closed, hot room with crowded seating is not conducive to a productive problem-solving session. Remember this is a team where all members are considered equal. The seating arrangement should reflect this. The best arrangement for the meeting should be a closed square/rectangle or a boardroom style using a round table. These two arrangements do not allow for "power seating." All are equal around the table.

A facilitator for the meeting would help keep the meeting on track. This person could come from within the team or outside. The facilitator would formulate the agenda, keep minutes, and keep things flowing, as well as evaluate the meeting to improve future meetings.

It is important to add here that meetings are no longer confined to a room. Internet and videoconferencing widen the opportunities for people to connect and meet with one another. The rules of communication are simply altered a bit when technology is the communication vehicle. For example, in two-way PictureTel® conferencing the person talking has a 30-second or so delay in getting a message to the listener (it seems to eliminate people interrupting each other and is a pleasant outcome of this method of meeting).

> **SOME BASIC RULES FOR AN EFFECTIVE MEETING**
>
> 1. Have an agenda of items to be discussed.
> 2. Send out an agenda and any materials to be reviewed two days before the meeting.
> 3. Have specific items to cover at the meeting. Limit these to only a few items.
> 4. Start and stop the meeting on time.
> 5. Record minutes of the meeting and send out copies to all team members.
> 6. All members should come prepared to participate.

Team Communications Should Be Organized

Whether your team communication is written or verbal, e-mail or posted, the following is a good format to follow:

- What — Explain the goal
- Who — Who is responsible?
- Why — The reason the task is being done
- How — Suggested actions that may help accomplish the goal
- Where — Location of where the job is to be done
- When — Project timetable
- What — Consequences, rewards, and penalties

Some call this the "3-1-3" method (www-h-www).

Leading the Team through Effective Communication

Team leadership is a fairly extensive discussion item but let's look at the role of the team leaders in terms of communication issues. The team leader, whether selected by the team or appointed by management, will be the team gatekeeper for any communication problems. The leader or coach will ensure that all voices are heard and will facilitate the consensus or voting process. In addition, the team leader will see to it that the regular written communications go out to the whole group, will conduct meetings, and will act as a liaison to outside teams. The meeting facilitator could also be the team leader or may not be.

The team leader may also take part in the confrontation of negative team behaviors. This role may not be comfortable but is necessary at times to keep the team on track.

> **TIPS FOR POSITIVE COMMUNICATION WHEN CONFRONTING NEGATIVE OR INAPPROPRIATE BEHAVIOR**
>
> - Always be honest.
> - Focus on behavior not the person.
> - Keep it team oriented — The problem is "our" problem because we operate in a team.
> - Keep the confrontation positive.
> - Focus confrontation as missed opportunities for the team rather than criticism of individual performance.

Often, confrontation can result in conflict. In the continually demanding, fast-paced work environment, conflict is bound to happen. Conflict is born out of passion for one's work; if people did not care about their work, they probably would not become agitated. But people in conflict can become allies instead of enemies. Here are some tips for managing conflict with others:

1. Validate others' feelings of frustration and anger.
2. Remain calm.
3. Don't get angry in return.
4. Don't talk loudly.
5. Build a bridge of commonality to stand on: "We are in this together and I understand."
6. Be clear on resolutions and who does what.

A conflict usually arises between two people when communication breaks down or there is a lack of understanding of another person's feelings. Managing conflict or differences in opinion can lead to a stronger team that makes better decisions and can promote diversity of thinking and creativity. To be creative, innovative, and to successfully problem solve, team members must also be open and receptive to frequent feedback. A group must learn how to compromise and be able to directly discuss the cause of problems. To keep the discussion from getting ugly, all team members must place the project goals on the table and check all personal feelings so they can be addressed. An undiscussed personal feeling is like a smoldering fire in a gas station. It is important to leave position and status outside the discussion. The issue is to focus on the job to be done, not who is in charge.

Listening Skills Complete the Communication Loop

When we are working together and trying to communicate, the biggest frustration comes when we know the other person is not listening. This type of communication is as effective as talking to a wall or as distorted as when we played "telephone" with a circle of people and the message came out totally different than it went in.

We tend to take listening for granted because we equate it with hearing. Hearing and listening are not the same. To effectively listen, there are practices that one can employ. This technique is referred to as *active listening*. Active listening is defined as the ability to hear a message clearly, using your eyes *and* ears, and completely concentrating.

EFFECTIVE TECHNIQUES FOR ACTIVE LISTENING

- Send listening signals
  - Make eye contact
  - Nod and acknowledge you have heard
- Minimize distractions
  - Focus on the speaker
  - Avoid fidgeting or pencil tapping
  - Avoid yawning
- Listen to the entire message
  - Avoid interrupting
  - Avoid giving your opinion until the end
- Reinforce the communication with verbal cues
  - "I would like to know more . . . "
  - "Elaborate on that last point . . . "
- Confirm the message
  - "I hear you saying that this is what took place . . . "
  - "If I understand you correctly, the following is . . . "

One of the biggest obstacles to communication is when we pass judgment on what we hear before we find out what the other person means. We must realize the other person's insights are real, even if they are not the same as ours.

Some other techniques for effective listening are to take notes and highlight key points. Mindmapping using diagrams and key words to capture the essential information is a useful tool. The effects of poor listening skills have been the cause of lost productivity, wasted time, and critical mistakes in the engineering, construction, and manufacturing process. This equal and vital part of the communication process has been long overlooked. If one focuses more of their effort on listening, total communication will improve. Active listening is one of the tools for building mutual respect among team members.

Communication Techniques to Use with Clients, Customers, and Suppliers

Our clients, customers, and suppliers are a vital component in the communication link between design, production, and delivery. If we look at points for communication breakdown, this is a critical point in the communication chain. If we don't hear the customer's needs or supplier's requirements, the product or project will fail to meet specifications. The skills used to communicate with customers and suppliers are very similar to communication skills in general, but some specific suggestions can help clarify customer or supplier expectations.

First, go in pairs to meet with the customer. Four ears are better than two and it is helpful to visit the customer sight firsthand to see the customer's operation or client's site for development. There is a tremendous amount of information a customer or supplier can share or contribute to the design process. Second, ask focused open-ended questions and solicit information about the customer's product(s), services, workforce, production schedule, and so on. Third, use your active listening skills and take notes. Let the clients or suppliers do the talking; you ask the questions and clarify points. Finally, when you are clarifying points, get specifics from the customer. When they talk "high quality," determine what this means in measurable standards.

> If you're talking, you aren't learning.
> —Lyndon B. Johnson

Sometimes you may hear negative feedback. Customers and suppliers can use you as a platform to vent their frustrations. Acknowledge them and try to redirect their communication to their needs. Specifically when working with suppliers, define customer needs with them. Communicate these expectations by clarifying the team objective for the use of their product or service and how it fits in with your team goals. Be clear on the expected performance in terms of delivery, quality, and costs. Ask the supplier to discuss their capabilities to meet your specifications. Once an agreement is reached, restate it for clarity and get it in writing. Then, establish a schedule for follow-up and review. If the service is working, give positive feedback to encourage continued performance. A process of feedback will help improve the relationship. Suppliers want to meet the company's needs but need to know how.

Negotiation Techniques

As teams and individuals work and function in large systems, negotiation becomes a tool of commerce. Negotiation techniques become critical in making a process work. Every day people wheel and deal on deadlines, supplies, use of space, use of personnel, and use of equipment. Knowing how to get what you or the team needs comes down to how to negotiate and influence. Here are some strategies to follow when negotiating.

- Have all your information. The more you know, the better the chances of convincing the other person or getting what you ask for. With information, you can provide solutions.
- Research the other side. When you are negotiating, know what the other person or team wants. What are their needs? Ask many questions. This information will help position your offer.
- Take time to consider an offer. Make sure the offer meets your needs before committing to it. Do more research.
- Never take the first offer. You can usually do better.
- Know when to walk away. Leave yourself room to walk away from a deal. Give yourself time to think. Sometimes it just isn't going to work.

- Cringe a little. Human nature will want to give you a better deal.
- Shoot high. Always ask for more so you have room to bargain. Both parties walk away feeling good.
- Lock in the deal up front. Once services are performed, it is always difficult to negotiate the deal.

As you negotiate and attempt to influence change, remember the following questions:

- What are the overall current perceptions of the issue at hand?
- What is your objective for changing or improving the situation?
- What will be the expected outcomes or performances?
- Do you have the authority to make a decision?
- Can you meet the expectations of the negotiation?

Most importantly, know your personal values and your team's values. Do not sacrifice these values in a deal.

Presenting Your Design: Making Your Case

After all has been said and done, the customer input has been taken, you've met with suppliers, worked out design glitches with your team, the time comes to present the final design. Where and who you present to is determined by the project and company procedures. It could be to a manufacturing team or to a community for public review. The processes for presentation are still the same. Today, many good computer software packages allow for clear, well-designed presentations. As the engineering field begins to communicate increasingly on-line, the need for good presentation skills will increase.

Every presentation should be supported with visuals that can emphasize key points. A handout with the same key points helps your audience follow along and make their own notes.

Match Your Visual Aid to Your Audience

There are general rules or guidelines for presenting visual materials. In small informal groups, white boards serve well to demonstrate uncomplicated materials. For more formal small groups, flip charts and transparencies are suitable. Limit the amount of information on each slide. Less is more in this situation. For large groups, large screen slides or computer PowerPoint® presentations work well. There are presentation units that attach to your computer to demonstrate software and even white boards, called Smart Boards, that project large screen applications of a software program. Although there are many options available, some basic rules still apply:

- Vary the delivery tool — audiences can be lulled into boredom.
- Limit your presentation time and allow for breaks or stretches.
- Check out the equipment before your presentation.
- Keep your visuals simple.
- Add color for impact.
- Organize in advance and determine your pace.
- Use handouts to emphasize important material.
- Don't hand out material until you want it read.

Putting it all together takes time. Allow enough preparation time to outline your presentation. Also know your material well because people will ask questions and more prep time directly translates into confidence at the podium. Remember that each presentation should have an opening, a central theme, some time for discussion, and a closing (see **Figure 5.8**) Work from note cards or allow your slides to trigger the dialogue. *Just don't read your presentation!*

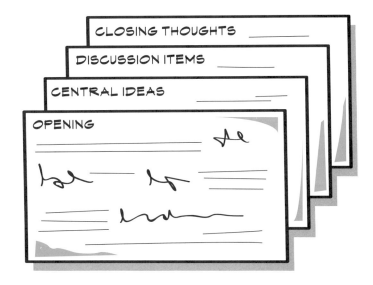

Figure 5.8
Presentations have an opening, central idea, discussion time, and closing

Presenting As a Team

Presenting as a team is common in the realm of technical operations. Architectural firms will send a team to present to a community meeting on a proposed project. Presenting in collaboration can be difficult but works well when it comes together. The key to making it work is assigning responsibilities for how it is to happen. Also, it works best when one team member is the coordinator. The division of duties should be based on the team members' strengths and respective roles in the project. Credit should go to all team members even if their work was behind the scenes. A dry run is advised so all agree on how to present and handle the discussion time. Success is based on total commitment to the process of the whole team from the start.

One final note on presenting your design: remember that first and last materials are what will be remembered. Use stories, humor, and real-life examples to emphasize your points. If there is one major point you want to convey, present it many times in several different ways.

Making Your Case

Often your team has a great project, has made all its deadlines and stayed in budget, but you are beat out for production by a competing team. What happened? Where did it go wrong?

Basically it comes down to selling and making a good case for your design. You must convince management that your design is the one that will bring in the big bucks, or can get the job done. This is called "do your homework." It is important to know the whole market or system within where your design can fit. The teams must have all the costs, data, and documentation to show how their design is better. Comparison data is critical and management wants to see both sides. The team must know their market, understand the customer, and point out the best features or selling points. The team also must be able to respond to the weaknesses. What solutions have the team come up with for these weak areas?

Once the team has made their case, they better be able to deliver. So honesty of the data process and proposal results is critical. If your team wins the proposal, there is nothing more frightening than winning on false data. The success of any project ultimately stands on real facts and figures and the ability to deliver the product on time.

If making your case fails, this is the time to meet with the team and go through a design review process in detail. The project may not be dead. With a review, changes can be made and next time it may be a success.

Summary . . . Links to the Whole

To conclude this section on communication and presentation skills, here are some tips to help you keep up with all the information and data that will impact you and your design project:

- Take responsibility for your part of the process
- Attend all informational meetings, virtual or real
- Locate all potential resources for information

- Become a power Internet user
- Read trade magazines and join professional societies
- Network with people knowledgeable in your specialty area
- Network, read, question, and ponder

Communicating a Design: Making Your Case

Communication is about people—ideas that are shared, thoughts, and feelings. Communication can be both positive and negative, but for it to serve a common purpose, effective tools must be employed. Listed here are many tools that work for effective communication to occur.

The more you can develop continuous awareness of the job, the stronger you will be in communicating the needs of the job to the whole system.

- Effective tools for corporate or group communication
    - Regular planned meetings
    - Protocol for meetings: parliamentary procedures, agendas, etc.
    - Newsletters
    - Weekly communication reports
    - Feedback sheets
    - Surveys and questionnaires
    - Assessments
    - Focus groups
    - Electronic chat rooms
    - E-mail
    - Groupware: Intranet/Extranets
    - Web sites
    - Videoconferencing
- Ask open-ended questions. These types of questions allow people to express themselves freely.
- Develop team norms that include communication as primary.

- Ten steps for reaching consensus
  1. Define the purpose of the decision to be made by the team.
  2. Use data and information to make intelligent decisions.
  3. Team members come prepared with their own feelings and data on the issue at hand.
  4. Team members share their thoughts about the issue.
  5. Listen to all team members' viewpoints.
  6. Clarify any unclear issues with questions.
  7. Value differing opinions, which help to look at the issue from a different point of view.
  8. Confront ideas, not people.
  9. Work for a quality decision that all on the team can support.
  10. Fully support the team discussion and individually take responsibility to see it through.
- Tools for building open communications — A continuous feedback loop: giving and receiving feedback.
  - Giving Feedback
    - Focus feedback to team results.
    - Base your perceptions on specific incidents and facts. Avoid opinions.
    - To avoid defensive positioning, involve the other person in the conversation.
    - Develop a plan of action that is jointly agreed upon.
    - Summarize the communication and the plan of action.
    - Let other people know you appreciate their participation and openness.
  - Receiving Feedback
    - Listen carefully to the major areas being addressed.
    - View the feedback as an opportunity for growth.
    - Ask for clarification by asking questions and paraphrasing key points.
    - Prove your point of view of the situation while remaining objective and avoid making excuses.
    - Offer your ideas for how to improve or change the situation.
    - Schedule a follow-up meeting to give a progress report.
    - Thank the person for wanting to help you by giving you feedback.
- Running an effective meeting increases your communications. Some basic rules for an effective meeting are:
  - Have an agenda of items to be discussed.
  - Send out agenda and any materials to be reviewed two days before meeting.
  - Have specific items to cover at the meeting. Limit to only a few items.
  - Start and stop the meeting on time.
  - Record minutes of the meeting and send out copies to all team members.
  - All members should come prepared to participate.
- Tips for the first team meeting
  - Introduce all members.

- Start on time.
- Choose a team name to build identity.
- Establish meeting times and dates.
- Choose a team meeting facilitator.
- Discuss the team leader position and choose a leader.
- Discuss team goals and objectives.
- Celebrate the new team.

◉ Tools for positive communication when confronting negative or inappropriate behavior
- Always be honest.
- Focus on the behavior, not the person.
- Keep it team-oriented — The problem is "our" problem because we operate in a team.
- Keep the confrontation on a positive focus.
- Focus confrontation as missed opportunities for the team rather than the criticisms of individual performance.

◉ Active listening is a tool to build mutual respect among team members. Effective techniques for active listening are:
- Send listening signals
  - Make eye contact
  - Nod and acknowledge you have heard
- Minimize distractions
  - Focus on the speaker
  - Avoid fidgeting or pencil tapping
  - Avoid yawning
- Listen to the entire message
  - Avoid interrupting
  - Avoid giving your opinion until the end
- Reinforce the communication with verbal cues
  - "I would like to know more . . . "
  - "Could you elaborate on that last point? . . . "
- Confirm the message
  - "I hear you saying that this is what took place . . . "
  - "If I understand you correctly, the following is . . . "

◉ Rules for good presentations
- Vary the delivery tool; audiences are often lulled into boredom.
- Limit your presentation time and allow for breaks or stretches.
- Check out the equipment before your presentation.
- Keep your visuals simple.
- Add color for impact.
- Organize in advance and determine your pace.
- Use handouts to emphasize important material.
- Don't hand out material until you want it read.

- Making your case
  - Know your competition.
  - Know your market and customer needs.
  - Document and collect data.
  - Know the selling points.
  - Provide solutions for obstacles.

A-Team Scenario

Presenting the Design Package to the Financial Group

"Hey, guys. While you all have been messing around with percentages, tolerances, and other such stuff, I've been fretting over here with our presentation package for our market feasibility materials. It needs to go out and I need your help!" Russell pleaded. "Sorry, Russell, we all have gotten caught up in the test of the pump. We've all been so busy we've not had our weekly meetings. I think that the team has gotten off schedule," Claudia responded. "What can we all do to help?"

"Well, I guess first we need to figure out what is needed in the report. Then we need to decide how we want to present the material," Russell explained. "Are we presenting in person or just sending the materials in to be reviewed by the marketing analysis group?" Shantel answers, "We are presenting in person. I have us set to meet with them next Monday." "Gee, thanks for letting us all know," the group retorted. "When did you plan for us to get this all done?" Carlos snapped. "Where is Joe? He's never here when we need to talk."

Tensions are running a little high now in the team as the deadlines get tighter. Communication systems have broken down. The group is finding fault with Shantel. Russell is feeling left out. Joe always seems to be gone. After a temporary breather, the team regroups and has one of their planning meetings.

At first the discussion was awkward. Each person was frustrated and in some cases angry. Claudia finally broke the silence, "You know we've been working awfully hard these last several months. Maybe we need a break." "I don't think so! We'd never make our deadline," Shantel replied. "Maybe we just need to talk it out," Russell countered. "Get what's on our chest out on the table." So the team did just that by reviewing the events of the last several months. Each person had their chance to talk and all team members agreed to listen. They did have some varying opinions. For example, Joe really didn't understand why it was necessary for him to be there all the time. He did finally agree to change his thinking to make it work for the team.

What they realized is they had just experienced a storming session. They discussed what had happened and made apologies to those who needed them. They then revisited their original list of norms and added a new one: all vital information will be communicated to all team members through e-mail. After that, communications went a lot smoother. To get the presentation ready for their Monday deadline, they all agreed to put in a couple extra hours a day to do their part.

"Carlos, what a great job you did in explaining the models of the pump and all the various levels of the design we can generate," Claudia exclaimed. "I think our presentation went very well. We all came across as a 'well-oiled team.' Shantel, your charts and diagrams really made everything make sense." "Yeah for PowerPoint," Shantel replied. "Joe, your analysis of the pump operation was well explained. I think the marketing group has enough information to do their feasibility study." Russell was just relieved it was over and he thanked his team for all their support and pitching in at the last minute.

Chapter 5 | Communicating a Design: Making Your Case

A-Team Scenario
continued

Respond to the A-Team scenario by answering the following questions. Use any tools or ideas you have learned in this chapter.

1. Why are regular meetings of the team so important?
   _____
   _____
   _____
   _____
   _____

2. List four rules for running an effective meeting.
   _____
   _____
   _____
   _____
   _____

3. Based on information related to working with a client found in this chapter, write two questions the team should have asked Earl, their client.
   _____
   _____
   _____
   _____
   _____

4. If you were putting the presentation package together for the marketing group, what tools would you have used to sell the pump design?
   _____
   _____
   _____
   _____
   _____

5. What activities indicated the team had moved into the forming stage? (Review team dynamics of Norming, Storming, and Forming in Chapter 2.)
   _____
   _____
   _____
   _____

TERMINOLOGY & DEFINITIONS

active listening — the ability to hear a message clearly using your eyes, ears, and complete concentration

agenda — outline of a meeting that details the objectives or decisions to be reached, as well as the location, date, and start/finish items

charter — a formal document that is a set of guidelines by which a group or organization must live

Activities and Projects

- **Answer** the following questions using your project as a focal point. These questions will help you and your team organize a presentation.

Determine the purpose of the project.

_____
_____
_____
_____
_____

Identify the audience (i.e., client, vendor, and/or upper management).

_____
_____
_____
_____
_____

Clarify desired outcomes.

_____
_____
_____
_____
_____

Chapter 5 | Communicating a Design: Making Your Case

Activities and Projects (continued)

Determine which parts of the project will be presented. Determine who will present each part.

_____
_____
_____
_____
_____

Choose methods of presentation (flip charts, white board, PowerPoint slides, video, etc.)

_____
_____
_____
_____
_____

Stage the presentation

Opening lines

_____
_____
_____
_____
_____

Closing thoughts

_____
_____
_____
_____
_____
_____

List all creditors. Remember the ones behind the scenes.

_____
_____
_____
_____
_____

Activities and Projects *continued*

- **Choose** one of the presentation methods presented in this chapter with which you are least familiar. Research the best ways to use this method for presenting your ideas and limitations of the method. Present your findings to your teammates. If this is done as a group assignment, have each team member choose a different presentation method.

- **Examine** and **assess** your communication skills or the teams' corporate communication skills.

    People skills:
    _____
    _____
    _____
    _____

    Organizational skills:
    _____
    _____
    _____
    _____

    Presentation skills:
    _____
    _____
    _____
    _____

    Research skills:
    _____
    _____
    _____
    _____

- **Apply** this to how you or your team will communicate and present the project.
    _____
    _____
    _____
    _____
    _____
    _____

Factors That Are Changing the Design Process

CHAPTER 6

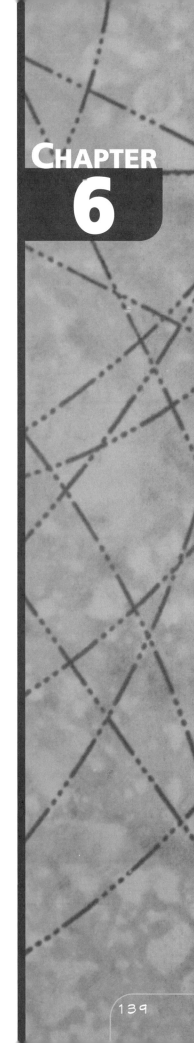

> "*I am enthusiastic over humanity's extraordinary and sometimes very timely ingenuities. If you are in a shipwreck and all the lifeboats are gone, a piano top buoyant enough to keep you afloat may come along and make a fortuitous life preserver. That is not to say, though, that the best way to design a life preserver is in the form of a piano top. I think we are clinging to a great many piano tops in accepting yesterday's fortuitous contrivings as constituting the only means for solving a given problem.*"
>
> R. Buckminster Fuller

Introduction

Human beings create systems to process and assimilate existing information to increase their ability to make things better, quicker, more precise, or more powerful. As we look at the evolution of the human condition, our knowledge of available resources and how to utilize them has only increased. The basic steps in the problem-solving process for design and ways of engineering have not changed drastically. What *has* changed is the structure and size of the engineering process system and the technology to deliver it.

Design systems today emphasize process. Traditionally, an engineering process has been a collection of activities that takes one or more kinds of input and creates an output that is of value to a customer. The delivery of the product to the customer is the value that the process creates. This is still true, but how the process works today is significantly different from the development and design of a product earlier in this century. If we look at Adam Smith's notion of breaking work into its simplest tasks, this is where the roots of traditional manufacturing and design can be found. Companies have used Adam Smith's model that focuses on individual tasks in the process of designing a product for a customer (see Figure 6.1). Engineering, testing, manufacturing, marketing, cost analysis, and so on, have all been part of a design cycle, but there has been limited communication between these groups. Often the larger objective of customer satisfaction and development of a quality product was lost with the individual focus on the task at hand. It is not to say the individual tasks are unimportant. None of them matter for customers if the overall process doesn't deliver the goods according to their specifications.

Figure 6.1
Adam Smith's view of manufacturing

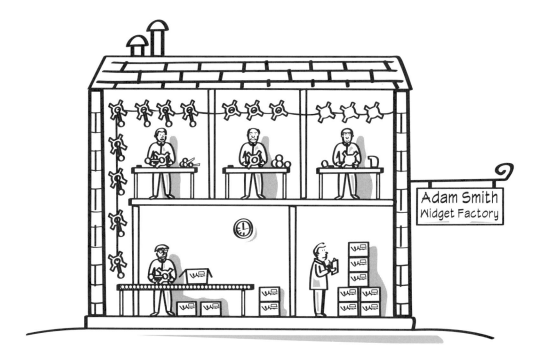

Today the design focus and process has changed. There are major factors that are bringing these changes about. This chapter will look at the factors that are changing the way designers are doing business when it comes to the design process and product development.

Better, Cheaper, and Faster

The design process is driven by the need for a better, cheaper, and faster-to-market product (see Figure 6.2). Global competition has accelerated to a point that the traditional design methods are not competitive. Communication technologies and the speed and volumes at which information is shared have accelerated the flow of activity at every level of society. This compression of time has impacted the engineering process and is requiring a different paradigm for managing the design process. Information technology is at the center of this change.

Figure 6.2
The design process is driven by the need for a better, cheaper, and faster-to-market product

As the formation of cross-functional teams merge, the ability to make decisions quickly is now possible. Access to all the expertise and information needed for decisions has been brought to a horizontal plane. The corporate structure of decisions coming from the top is breaking down. A decision is now as close as the desktop computer and an Internet chat room with team members (see **Figure 6.3**). The speed of decision-making is only limited by the speed and capability of the computer.

Figure 6.3
Information technology is the link to the changes in the design process

The design cycle is shortening so the product can get to market faster. In many cases, being second on the market dramatically drops the profit margin for the product. Life cycles of product designs are also shorter, so the ability to respond to the needs of the customer has been a critical factor for corporate survival. It is predicted that the design cycle will even speed up more in years to come. We are just at the beginning of this new paradigm. *Design News,* surveyed its readers in 1999 and 84 percent of their readers expect shorter design cycles in the next five years (see **Figure 6.4**).

Figure 6.4
Dawning of a new paradigm

So, how do we design and produce products faster? Listed here are ways the industry is responding to this phenomenon. We will give a closer look at the factors that are changing the way we design and the implications to the designer.

- The customer is at the top of the organizational chart.
- Cross-functional teams
- Virtual teams
- Virtual design environment
- Design for manufacturability (DFM)
- CAE/CAD/CAM software: 3D modeling and what it means for design
- Product data management (PDM)
- **Rapid prototyping**

The Customer is at the Top of the Organizational Chart

When we think of designing a product or project, usually the first focus for the designer is the product itself. With the focus on design, the design process becomes isolated and removed from other steps in the overall system. Changes evolve from within the design. Today the focus has shifted from the specifications being driven by the product design to the needs and specifications required by the client or customer (see **Figure 6.5**). One United States Company, Drake Manufacturing, reports that there is a 100 percent direct open line between their customers and engineering. Their personnel are expected to make six customer visits per quarter. The designers follow the product from creation to installation and service, since Drake does not have a service department. Of the forty-five employees at Drake Manufacturing, thirty are involved with the customers (*Tooling and Production,* May 1998).

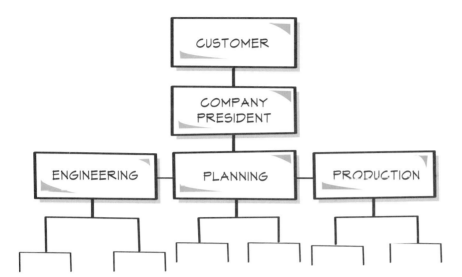

Figure 6.5
The customer is at the top of the organizational chart

Another example is Rocky Mountain Consultants, a civil engineering firm in Longmont, Colorado. They describe a project for land development in a small town in Colorado. A local farm family deeded over their land to the town of Kersey. The land development process began with a town meeting. The final development plan reflected many of the requests from the community that were generated at that town meeting, including the ballpark the children requested. Regular meetings were held at various points in the design process to keep the community informed and to solicit any changes that would affect the design.

This focus on customer satisfaction and customer specifications eliminates the chance for product failure, builds happy clients, and a solid working relationship. This focus on the customer speeds up the design process by shortening the amount of design changes and reworks on the manufacturing floor or at the construction site.

Implications for the designer

1. Good communication skills
2. Good listening skills
3. Understanding of customer's product/client's project

Cross-functional Teams

We have studied teams and their development in Chapter 2, Teams as a Tool in the Engineering Design Process. Here we ask, "Why do teams work well in a design cycle?" Since teams bring together all the expertise of the organization, today's challenges require pulling and working together across functional and organizational barriers. It's a funny thing — put a group of people together no matter their differences, hand them a challenge with a deadline and they can perform miracles. Industry has many success stories of how teams have redesigned workflows to make them more performance driven and effective.

The boundaries that have existed between different functional departments that are key to the design cycle in the past have slowed progress. By bringing every one together with a common goal, you build the synergy of a team.

INDIVIDUAL VERSUS TEAMWORK MODEL

THEN	NOW
Direction given by management	Team determines its own direction based on process needs
Separation of workers and designers	All team members contribute to the process
Individual accountability	Mutual support and joint accountability
Individual doing one task well isolated from the whole process	Team members play multiple role and work together towards continuous improvement

Changes in the workforce are required by changes in the design cycle.

Implications for the designer

1. Trained in team dynamics
2. Content expert
3. Good communication skills
4. Knowledge of the whole system and the related functions

Virtual Teams

The concept of systems engineering is an efficient way to solve problems of communication between diverse groups. Product or project development teams now have at their disposal computer access that allows them to build common understanding and communication regarding the development of a project. In the traditional model of design and manufacturing, a limited number of people held the knowledge of all the concurrent processes. This is not possible today because of the complexity of the product production systems and because of the speed now required for development. One person cannot keep track of the whole design cycle. Now we need a team of experts that can communicate, problem-solve, and create quickly and efficiently.

Industry Scenario

E-Manufacturing

— Richard E. Neal, President, IMTI Inc.

E-commerce is revolutionizing the way business does business. In the United States, e-commerce was a $507 billion industry in 1999 (http://www.internetindicators.com/); by 2003, this number will top $1.7 trillion ("Resizing On-line Business Trade," Forrester Research, Inc., 1998). While much of the visibility is on business-to-consumer e-commerce (like e-bay and amazon.com), the value of business-to-business commerce (exchanges between business entities) is five times as great!

There is much more to the e-commerce equation than just buying and selling. E-commerce demands a totally connected environment. It demands a level of security and confidence that doesn't exist in today's systems. It also points to new opportunities. Chief among them is to create a totally electronic design-to-manufacturing environment — let's call it e-manufacturing. If we are to truly realize the vision of e-commerce, we must achieve the ability to electronically manage the complete product realization process in a totally connected (all systems talk to each other), inter-operable (all systems plug and play), seamlessly integrated (no breaks in the information flow) environment. The product realization process spans the total manufacturing cycle, from the inception of the idea, to delivery of the product and process design, through the manufacturing process, to the end user.

While e-manufacturing presents many technological challenges, it can well be illustrated with an example. An automobile manufacturer must redesign an engine to meet emission standards, without compromising engine performance. Through simulations, they determine that the changes in the emission system demand increasing the engine horsepower. The requirements are input to the system, and options are evaluated. Through an interactive process, the data is presented to the design and manufacturing team with the recommendation that the cylinder bore be increased. However, the company wants to avoid the time and cost of redesign and retooling, so the best-balanced alternative must be found. The design and manufacturing team, "virtually co-located: (located anywhere but connected via the Internet), inputs design alternatives. The design system (a "virtual cockpit") launches simulation tools through an optimization engine that determines the best options for bore size and location while still protecting engine life, thermal properties, and stress concerns. Unlike today's systems, all of the simulation tools are integrated, so all of the parameters (including life-cycle issues) in the total process are included in a single simulation system. The design and manufacturing team can thus select the final design based on a full knowledge of all options and trade-offs.

Once the design is chosen, the system automatically updates the master product design with the new parameters. The system then launches the detailed manufacturing processes design function whereby science-based knowledge systems generate all the resource and control information necessary to make the new engine components. Enterprise-wide manufacturing management systems assure that the right information is provided to the right machines, and the new parts are made in an environment that assures product quality through intelligent control of all processes.

E-manufacturing is an emerging factor in the e-commerce equation.

This team of experts comes from different locations and represents different fields of expertise. In systems engineering, the design process is not sequential. The process is interactive and occurs in phases that are interrelated. The function of this team of experts is to make decisions to keep the process going by using the information that is available. For teams, the communication method today has become virtual. It is not necessary to meet in a conference room, though this is still part of the connecting process. Today, teams work together in cyberspace (see **Figure 6.6**).

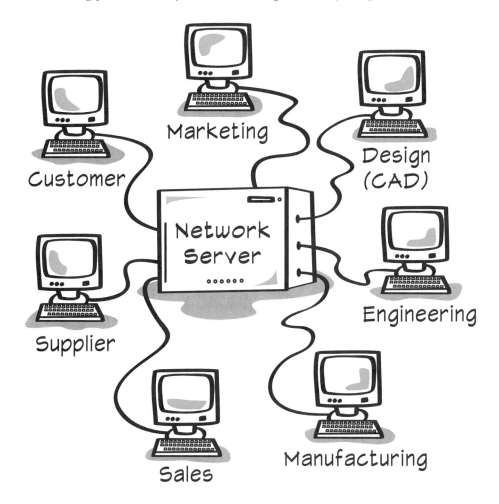

Figure 6.6
Virtual teams

Virtual Design Process

A virtual team has the responsibility of input and access to a database to which all team members have access. A common server hosts databases and tools needed in product development and communication. The user interface is a common virtual environment. This is a place where the team can meet each other and examine the 3D model of the product or documents. Discussion areas (i.e., on-line meetings and videoconferencing) and e-mail applications will aid in asynchronous communication. Companies are developing knowledge archives on competitive information, customer research, and lessons learned from other projects. The archive is a library of accessible knowledge to help in the design development process. It receives data input and output from all team members.

The virtual interface could handle or facilitate such processes as handling customer inquiries, creating a concept for a product, bid preparation, design for assembly, design review, and storage, shipping, and billing.

This virtual interface allows for a virtual design process. The 3D CAD/CAM/CAE systems that are currently available allow designers to model, evaluate, test, and refine complex engineering problems without the costly process of developing prototypes. The traditional model for developing a design was to first prepare drawings and then build a physical prototype (see Figure 6.7). Just imagine all the wiring in a space shuttle. In the design process, the designer or engineer had to physically create prototypes of wire harnesses and cables that connected all the electronics in the shuttlecraft. With tape and string in hand, each wire length was measured to guarantee the correct length for each connection. The time to accomplish this must have been a huge chunk of precious design cycle time, as well as questionable in terms of accuracy.

Figure 6.7
Traditional design model

Today, this is done completely in a virtual environment within a 3D CAD model. The development cycle is accelerated and the results improve accuracy. Documentation such as bill of materials, wire run lists, and so on are automatically produced. This documentation is associative in that as the CAD model changes, the documentation updates itself. With the new level of CAD products, reports can be produced from data stored in the system (see Figure 6.8). This data can be used to analyze and optimize the design. Weights can be calculated, which helps in design decisions for the product. The bottom line cost data is now available for the product design.

Design Tools for Engineering Teams

Figure 6.8
Today's design model

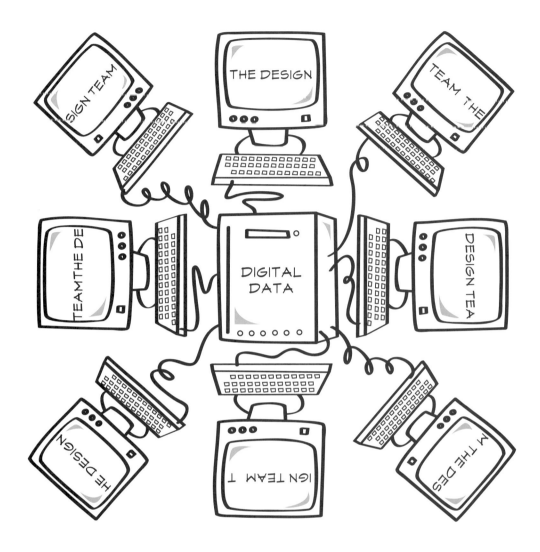

This is another virtual tool to maximize product design while minimizing costs. Because all this data is in digital form, it is accessible to any member of the virtual team. In addition, as the development of the design progresses, the data automatically is updated as new information or design changes are implemented to the design model. Even now, virtual manufacturing software packages are being utilized to plan plant operations. A system can be designed, evaluated, and the operations optimized before the facility is even built. Once built, modifications can be tested before implementation without disruption to production.

The virtual environment allows for a more complete and comprehensive plan for a product or project development. The virtual designs have reduced time to market by eliminating unwanted iterations.

Chapter 6 | Factors That Are Changing the Design Process

ADVANTAGES OF VIRTUAL PROTOTYPING SOFTWARE

- A system review enables you to examine the impact on the overall product.
- Because it is integrated, it concurrently analyzes the trade-offs between the product/project's costs, size, manufacturability, and performance.
- Because it is highly accurate, product development is quoted within 2 to 8 percent accuracy,
- Virtual systems support many technologies from design modeling and packaging to the bid sheet from the vendors.

The virtual design process is being pushed along by the auto industry with their need to shorten the time to market while lowering development costs. In this virtual process, quality is the outcome. As businesses ask themselves why they should go virtual, those who already have boast of faster market response time and lower production costs. The market has shown that 80 percent of the cost and performance of a product is established in the first 20 percent of the design cycle.

The design cycle now requires bringing design and implementation discussions to the front of the process. Misinformed decisions made in the early stages of the design cycle will result in higher costs, lower performance of the products, and missed time-to-market targets. This means if the design team has tolerance problems with a part, then they will likely have problems in production as well. The virtual environment allows the design team to simulate the design and production processes up front so later on the tolerance problems have already been worked out.

If we look at the four basic stages of the design cycle — dream, design, develop, and produce — and relate costs of making changes, in the development phase these costs are twenty times greater than the cost of change during the dream or conceptualization stage (see **Figure 6.9**). With the development of effective computer simulations, this is possible now. The cost of changes in the production stage can be 100 times the cost of the change made in a simulation model.

> "We lost $10,000,000 in revenue because we had a two-month delay to do another design iteration because we found out too late that the design wouldn't fit."
>
> — A frustrated anonymous author

Figure 6.9
Cost of design changes

The designer's success on the design team depends on an ability to manage and manipulate advanced technologies such as CAD while minimizing and reducing costly design iterations, as well as the ability to understand each part of the whole system as it relates to product development.

Implications for the designer

1. High skill level with computers
2. Knowledge of working with data
3. Good decision-making skills

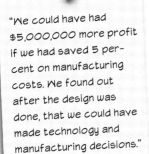

Design for Manufacturability (DFM)

Design for manufacturability (DFM) is just one of the commonly used names for a quality improvement process. Other terms for this same process are simultaneous engineering, concurrent engineering, design for assembly, design for competitive advantage, and design for excellence. No matter what the name, the approach is integrated and takes into account the factors that impact quality, customer satisfaction, and project implementation or product production. The main impact of DFM on the design process is that the manufacturing capabilities are considered as the product is designed. This insures that the product can indeed be made and reduces expensive design rework later in the design process. The DFM process has the customer first with quality as its team partner. The end result is a higher quality product at a lower cost.

> "We could have had $5,000,000 more profit if we had saved 5 percent on manufacturing costs. We found out after the design was done, that we could have made technology and manufacturing decisions."
>
> —Anonymous

Implications for the designer

1. Knowledge of manufacturing systems
2. Knowledge of quality standards and how to measure them

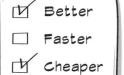

CAD Software: 3D Modeling and What it Means for Design

Computer-aided drafting and design (CAD) has long been an important design tool for technical and engineering applications. Primarily it has been utilized in 2D drawing applications even though 3D applications have been available. In recent times, 3D modeling software has become an affordable choice for CAD users. As 3D modeling has become more accessible, the way to utilize it has expanded. For developing a product design virtually, 3D modeling is a powerful tool. Its power lies in its ability to create concept drawings in the preliminary stages as well as in producing drawings for the final product.

It is hard to convey ideas as effectively as a 3D model can. The 3D model can show the whole assembly or mechanisms and how each section works. The 3D solid model allows for management of assemblies that are composed of multiple modeled parts and manages complex constraint hierarchy. As parts are being designed and built in the computer, many changes occur. The 3D model allows for one change in the model to dynamically change all other parts that are related. In addition, a history of command inputs is held in the database if a designer needs to retrace the development steps. Assembly complexity makes design in a 2D format difficult, even impractical now. A shipbuilder can utilize 3D modeling to obtain a clear picture of the entire ship. The client who commissioned the ship is able to view what they had ordered. In this example, 3D modeling is extremely helpful in defining where the piping and equipment exists so these can be prebuilt in sections or subassemblies and brought together for assembly in one unit.

When working with assemblies with thousands of parts there is a difficulty in checking each dimension and the interfaces. With assembly modeling in CAD there is a guarantee that everything fits together before the first part or prototype is built. One industry example states that the use of 3D solid modeling to check fit and interferences shortened the development cycle by one year (see **Figure 6.10**).

Figure 6.10
CAD modeling and more

Simulations

Finite Element Analysis (FEA)

Another design tool available in CAD is finite element analysis (FEA). Now designers can analyze products for weak areas through the available CAD software that has FEA in its software package. These weaknesses can be corrected virtually in the early design stages. Typically, the designer will test a product until it breaks to find weaknesses. This was a time-consuming process and has become known as destructive testing.

FEA is a tool that helps designers to get closer to the best design, and allows them to go through fewer destructive tests. As a result, the design cycle has been shortened considerably.

Most FEA applications are used to evaluate static pressures and deflections, but can also be used for dynamic, thermal, and fluid analysis. Finite element modeling (FEM) is the part of the software that prepares the model for analysis. In FEA, this is embedded in the program.

Kinematics and Dynamic Software

To model moving parts, kinematics and dynamic software is used. Kinematics is the study of motion without regard to forces. This type of software typically checks for clearances and interfaces. CAD programs are able to create dynamic simulations to study moving assemblies. The animations show the interaction of all the parts and the velocity of the moving parts is charted in graph form. To create a simulation, the geometry of both parts and how the parts are free to move must be shown and defined. In today's CAD software, this is automatically captured during the geometric modeling and definition of constraints.

Rendering Improves the Design Engineering Process

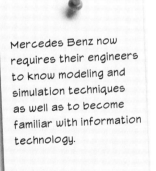

Mercedes Benz now requires their engineers to know modeling and simulation techniques as well as to become familiar with information technology.

The main point of rendering is the encouragement of communication and understanding of the part or assembly model. They say "a picture is worth a thousand words." We might want to reword that and say "a picture sells a thousands designs." The rendered designs have a lifelike appearance in terms of color, material, light reflection, and refraction. The rendering allows the designer to show designs in their intended materials. A designer may change the class value of adjacent parts that might otherwise become obscure to each other, a handy tool to facilitate assembly and field service. Rendering is also a productive tool for designers to see imperfections in complex surfaces. This allows them to be corrected before creating a rapid prototype.

Product Data Management (PDM)

Product data management (PDM) or product information management (PIM) is used to control information, files, documents, and work processes needed for the development, design, support, and distribution of a product. Product management data usually includes the following:

- Design geometry
- Engineering drawings
- Project plans
- Part/data files
- Assembly drawings
- Product specifications
- Numerical control machine tool programs
- Engineering change notices
- Tool library information

As we consider the virtual team and the ability to access this data, we can find examples of how companies are using intranets (internal company networks) and extranets (intranets to outside select companies) to communicate and collaborate with each other as well as to the customer and suppliers. Some are using Web sites that allow users to access many different sources of data and display the information. The Web browsers eliminate the need to install and maintain specialized software on each station. A Web browser can access PDM data. In addition, it can tie into ERP systems (enterprise resource planning) that give data on parts management, purchasing, finance, and shipping. The Web browser has become a link that allows the whole system to connect at least from a voyeur's point of view (see Figure 6.11).

Figure 6.11
PDM links

> **HOW CAD AND WEB SITES WORK TOGETHER**
>
> - CAD designs can be securely published on project-specific Web sites.
> - Web documents use a file format called HTML (Hypertext Markup Language). This language allows the Webmaster to control Web page design.
> - DWF (Drawing Web Format) is a file format that allows CAD data to be embedded into HTML files. (DWF was developed by AutoDesk.)
> - "Whip" is a driver to help Web browsers to utilize DWF files. ("Whip" was developed by AutoDesk.)
> - VRML (Virtual Reality Modeling Language) are images that are 3D VRML in order to function on the Internet.

PDM is the integration tool that connects all product data. As PDM systems have been developed, the CAD vendors have chosen one common avenue of communication: the Internet. The Internet has become the information technology for engineering firms to connect and transmit data and documents.

> **PRODUCT DATA MANAGEMENT (PDM) FUNCTIONS**
>
> - Controls data storage and retrieval
> - Searches for data by user-defined attributes
> - Coordinates relationships between different kinds of data
> - Manages check-in/check-out files
> - Tracks engineering changes and approvals
> - Maintains proper design revisions
> - Converts and synchronizes the flow of data between different applications
> - Provides product configuration management
> - Generates part numbers and other identifications

Another advantage of the 3D modeler is the ability to allow the designer to quickly generate cross sections of complicated areas in the assembly. A 2D view through an area can still be a useful tool to understand part interaction.

Let's recap this section by looking at improvements the changes in CAD have brought to the design process:

There has been an improvement in the graphic representation of the designed object. With CAD, the model can appear on the screen like a realistic image, in movement, capable of being observed from different perspectives. Whenever required, a printing device, a plotter, will provide a paper copy of a view of the geometric model.

There has also been an improvement in the design process. It is possible to visualize details of the model, to check collisions between the parts, to query distances, weights, inertia, and so on. In short, the process of creating a new product is optimized, while lowering costs, gaining in quality, and reducing the time involved in design.

The improvements for the users include higher productivity in the sketches of plans, integration with other stages of the design, more flexibility, and greater ease in making modifications to the design. Moreover, CAD facilitates standardization, enables a reduction in the number of revisions and provides better control of the design process.

Although the utilization of CAD in the automobile and equipment goods industries is remarkable in the mechanical design of parts and/or machines, there are other industrial sectors using the CAD technology, too. It is used for the electronic design of circuits (CAD-2D), architecture and civil engineering, industrial engineering (offices and industrial factories, town planning), for pattern making in the textile industry (CAD-2D) and in many other industries such as the graphic arts and animation industries.

Implications for the designer

1. Extensive knowledge and applied skills for CAD systems
2. 3D visualization skills
3. Data management skills

☑ Better
☑ Faster
☑ Cheaper

Rapid Prototyping

Designers and engineers may go through a thousand iterations in a 3D CAD model before the first prototype is built. The trade-off with the computer is computer time versus shop time. There is another computer tool that is impacting the design cycle. The technology is referred to as *rapid prototyping*. Through 3D solid model CAD software, mathematical data for solid 3D parts are created. These computer images are transferred into mathematical data that is transferred to a rapid prototyping machine (see Figure 6-12).

Design Tools for Engineering Teams

Figure 6.12 Stereo lithography process

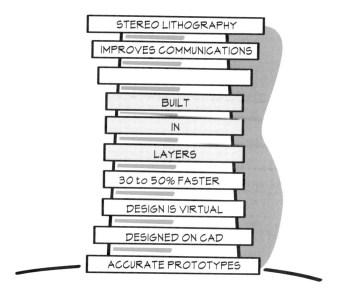

Benefits of a Rapid Prototype Model

Rapid prototype development provides many advantages to the design cycle. Designers can go from early design to an actual virtual product, and since they have a physical part to evaluate and critique, this allows for feedback from customers before final production and tooling. Designers can also create accurate prototypes from part data and ensure that parts are built 30 to 50 percent faster. Rapid prototyping also serves as a digital communication tool to teams and improves communication about the development of the part. This, in turn, speeds up the design development time and decreases design time and cost.

Rapid Product Development Leads to Rapid Tooling

Motorola, Inc. is a current leader in the development of both stereo lithography (SL), which is one type of a rapid development modeling tool, and rapid prototyping tools. They benchmarked stereo lithography (SL) in 1989 for 3D systems and it has taken all this time for it to move into the forefront of the design process. One reason for it to become more generally utilized as a design tool is the affordability of 3D solid modeling software.

As 3D solid model software paved the way for SL, SL and other rapid product development tools are paving the way for rapid tooling and quick-casting processes. In the traditional model, the product design went to the machine shop. Dies were machined for injection molds. The process has been lengthy and expensive. The dies used for investment casting operations were machined from blocks of metal that produced cavities used to produce wax patterns.

The rapid tooling design process steps are as follows:

1. The part is designed in the positive on a solid modeling CAD system.
2. The positive of the part is built using rapid product development tool.
3. The design is evaluated, verified, and iterated.
4. A negative of the part is generated in CAD.
5. Tooling personnel determine parting planes, draft angles, fillets registration holes, and ejector holes for the part.
6. Divide the negative part image into a core and cavity tooling pair.
7. Build a tooling pair in quick cast.
8. Investment cast the quick-cast tooling pair in the desired metal or injection mold in plastic, rubber, or nylon.
9. Perform all final machining.

Rapid tooling has saved time to a final product and improved quality because the design goes to rapid tooling as a mature design. Most changes or errors are caught in the CAD phases or even the quick-cast phase. Let it be noted, changes like material of choice or process of choice changes will not be found in the rapid product development or rapid tooling of a part. Material and/or process of choice changes often cannot be identified until a part has come off a production tool. Costs are reduced because it is cheaper to change a CAD design than it is to redesign later. Another example would be when the critical design elements occur early in the design development process.

Implications for the designer

1. Knowledge of total CAD systems
2. Dimensioning and tolerancing
3. Tooling processes

☑ Better
☑ Faster
☐ Cheaper

Industry Scenario

Only the Fast Survive

— *Jim Leonard, Colorado Manufacturing Competitiveness, Denver, Colorado*

Every morning in Africa a gazelle wakes up, knowing it must outrun the fastest lion or it will be killed. At the same time a lion wakes up, knowing it must run faster than the slowest gazelle or it will starve. It doesn't matter whether you're a gazelle or a lion, when the sun comes up you'd better be running. It's the law of the competitive jungle — only the fast survive.

Source: A gazelle somewhere in Africa

Every product development team member has heard or read the law of the competitive jungle; many have the law stamped on their foreheads. So, why is rapid product development, a.k.a., quick time to market, so important? There are many reasons, but the main one is: *Satisfy customers' needs first and best.*

Design Tools for Engineering Teams: An Integrated Approach describes many tools and practices that product developers use to improve product quality and

Industry Scenario, continued

reduce development time. But the toolbox of processes, practices, and technologies is constantly changing. What is state-of-the art today, will generally be mainstream in two or three years. State-of-the-art tools are rarely visible to anyone other than the most insightful industry observers because best-in-class product developers keep their new tools and practices secret as long as possible. Here is a brief list of practices and technologies that best-in-class teams are using right now.

Customer focus

The 1990s were characterized by the increased role of "customer focus," particularly leading-edge manufacturers. That leading role will intensify in the 2000s, as articulated by Jacques Nasser, CEO of Ford, " . . . the real work at hand; getting inside the mind of the consumer to understand what he or she wants, aspires to become, and will be needed long after a purchase has been made." Product development professionals will be employing all kinds of innovative ways to understand customer needs, wants, fantasies, including living with customers, having customers as permanent members of product development teams, drilling down into and analyzing mountains of consumer data collected over the Internet.

E-business

Current applications of the Internet provide a distinct competitive advantage. Although Internet applications are appearing from everywhere to improve customer service and reduce cycle times, most of the highest value-added applications are in product development. New, innovative applications will continue to appear. Every one of the practices mentioned in this brief list applies Internet technologies in some way.

Collaborative engineering

Collaborative engineering goes beyond concurrent engineering. It is the cooperative exchange of resources, for example, information, and ideas among a virtual team focused on an engineering-intensive project and having an overall common creative purpose. As you recall, a virtual team is one whose members are not physically collocated, but connected by distance communication technologies such as videoconferencing, and e-mail.

Predictive engineering, simulation, and virtual prototypes

With mainstream use of 3D solid modeling, products can now be designed, built, tested to failure, and redesigned all in digital form. The designs can be nearly optimized before time, funds, and effort are spent on hardware. Entire manufacturing plants can be simulated and optimized before the first cubic yard of concrete is poured.

Visual engineering

3D solid modeling also allows all team members to see and understand information in the same way. Now the marketing, finance, and other nontechnical members can see product designs without needing the skill to read 2D engineering drawings. Realistic images of engineered products can be shown quicker and for much less cost than building physical prototypes.

Supply chain integration, and value chain integration

Companies and teams now coordinate the series of activities and processes that purchase, design, manufacture, and deliver products and/or services to customers. The Supply Chain or Value Chain is a network of company relationships that support material, information, and funds flows from a firm's supplier's supplier to its customer's customer.

Design for everything (DfX) has new meaning

DfX used to mean something such as "design for manufacturability, assembly, test, service, and environment." Now DfX is taking on several new functions.

- Design for supply chain:
- Optimizes use of the distinctive competencies of all supply chain members
- Design for postponement:

> **Industry Scenario, continued**
>
> - Allows for incorporation of distinctive/customizable features into a product until the latest possible production step
> - Design for recycle, reuse, rebuild, and disposal: Design for environment used to mean "consider ease of recycling." In the future, expensive, high-information-content products will be designed for easy, inexpensive rebuild/refurbish, which eliminates the need for recycle or disposal for at least another life cycle. Less expensive products will be designed for convenient, environmentally acceptable disposal, which may include recycling or components made of biodegradable material.
>
> This list is certainly not comprehensive. Most of these practices and technologies may be mainstream by the time this book is purchased and read, certainly by the early 2000s. Enlightened product development professionals constantly scan the horizon for new practices, processes, and technologies, and then intelligently apply them in their own projects. This is the only way to consistently *satisfy customers' needs first and best*.

How the Design Process Is Responding to Technology

As we look back on the advances in technology that were covered in this chapter, we cannot overlook how this ultimately has changed the thinking and the processes for designing. Let's look again at the engineering process. There are several truths for design that are emerging in this new world of systems engineering.

- Timing of the design — do it early in the design cycle. Doing it later costs more.
- Reduce parts — reduces costs throughout the whole assembly process.
- Standardize parts — reduces cost by interchangeability of parts for different designs.
- Keep the design simple — the higher the tooling costs, the more complicated the design.
- Utilize modular designs — these reduce cost in design time because parts can be used in other designs. Modular designs also reduce assembly time.
- Design with gravity in mind — it is easier and time efficient to assemble from the top down.
- Eliminate fasteners — these save time and cost in assembly as well as end-of-life disassembly.
- Optimize part handling — costs go down and quality goes up with minimal proper assembly sequence.
- Design for easy part mating — easier assembly through alignment for insertion speeds up assembly.
- Provide nesting features in the design — these help to show where features go.
- Optimize manufacturing process sequence — this speeds up the product to market.
- Form follows function — design for the *user*.
- Utilize design generations — an initial design can be reused in the next generation of the product.

Dr. W. Edwards Deming, known for his work in quality management, stated that the sequencing of the work process for a design project would lead to a more efficiently produced product. He believed that deviation from the natural sequence produces rework and errors that cost time and money to fix. The best way to determine this sequence is to prepare a task diagram for each project, because the rules change for each project or product. A task diagram will allow the design team to evaluate and track design changes (see Figure 6.13). Design requirements not met in the early stages of a design will always add costs. An extreme example is with the design of the Corvette car. The suspension on one model was under designed and it went all the way through production. It cost $1,000 a car to retrofit a $10 part.

Figure 6.13
Task diagram example

©1997, Fred Pryor Seminars, PM-0977-US, pg. 9.

TASK OBJECTIVE

Task Milestones	Measurement	Time
1.		
2.		
3.		

Task(s) on which this task is dependent:

Task(s) that are dependent on this task:

Budget/Cost Factors:

How Recycling and Product Retirement Affect the Design

Another area that is currently impacting the design process is design for ease of recycling or design for disassembly (DFD). Designers today are being asked to design parts and assemblies so products can be easily disassembled at the end of the product's life cycle. Material recovery and disposability will influence the cost for the total design package — from the birth of the design through the active life until death. The impact to the design process is:

- Select recoverable materials for part or product design
- Design for easy disassembly — Design for Disassembly/Environment (DFD/E)
- Choose materials based on recycling costs
- Design in clumps. Clumps are groups of parts or components for which there is a common recycling process or reuse. The designer will compare various clumping strategies for a product design and choose the best option.

Summary . . . or Links to the Whole

Designers have a new way to view product development. It evolves within a whole systems approach and develops in a design chain. A design chain from conceptualization to product in hand, to product retirement. The process is integrated, responsive, front-end loaded, virtual, and continuous. One design chain can support a new one and develops from the knowledge gained from a previous design. The whole process is dynamic (see **Figure 6.14**).

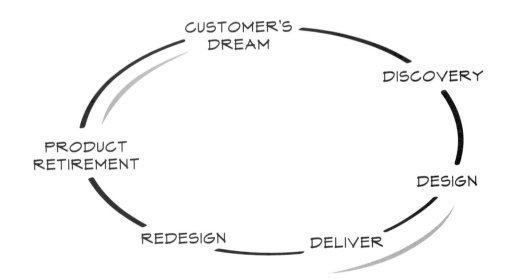

Figure 6.14
The design cycle

Factors That Are Changing the Design Process

Changes in the way we design have become a fact of life in engineering. These changes will continue to occur as the design community becomes familiar and skilled in new technologies. Listed here are current factors that impact how we design.

- Tools to produce designs and products faster and more accurately
  - Focus on customer needs
  - Cross-functional teams
  - Virtual teams
  - Virtual design environment
  - Design for manufacturability (DFM)
  - CAE/CAD/CAM
  - Product data management (PDM)
  - Rapid product development
- Virtual interface tool can:
  - Handle customer inquiries
  - Facilitate the conception of a product
  - Prepare bid preparation
  - Facilitate design for assembly
  - Facilitate the design review process
  - Establish storage, shipping, and billing
  - Aid in the transfer of data to the virtual team

Chapter 6 | Factors That Are Changing the Design Process

- There are five basic objectives and benefits that describe DFM
    1. Develop products that meet the customer quality expectations and specifications
    2. Design products with manufacturing capabilities in mind
    3. Reduce product development time
    4. Improved productivity and lower costs by producing designs that do not require extensive redesign work or engineering changes
    5. Produce a quality competitive product
- Five essentials for an exemplary integrated product development system
    1. CAE/CAD/CAM tools
    2. Cross-functional teams
    3. Good documentation for system management
    4. Strong program for integrating engineering and production
    5. System of early product definition, design reviews, and customer involvement
- Data used in product data management (PDM)
    - Design geometry
    - Engineering drawings
    - Project plans
    - Part/Data files
    - Assembly drawings
    - Product specifications
    - Numerical control machine tool programs
    - Engineering change notices
    - Tool library information
- Rapid product development tools/rapid tooling

A-Team Scenario

Factors That Have Made the Pump Design Easier

Today when the A-Team got into work, the answering machine was blinking. It was a message from Earl. He sounded a bit anxious on the recorder and asked for one of them to give him a call as soon as possible. Joe immediately took the message and returned his call. "Hi, Earl. This is Joe of the A-Team Design Group. What's up?" "Well, not much. Well, maybe there is something," replied Earl. "I just returned from a trade show for Ranch and Farm Equipment. I was snooping around to see if my pump idea would work, but what I heard is making me nervous. I think we have some competition. The rumor has it that some outfit in the Midwest is coming out with a version of a small pump in about six months. Are we going to be able to get our pump out when we said?" "Earl, let's not panic. We are pretty far along in the development of your pump. I'd say we are at least four months ahead of this guy in the Midwest. We realize 'first-to-market' gains the majority of the market share," responded Joe. "Have you located a foundry that can pour the casting for a pump this size?" Joe continued. "No, I guess I better get on that. Also, did you find out if we need a UL rating?" Earl queried. "What Shantel found out is that we only need to meet the ANSI specifications for pump designs," Joe stated. Earl asked, "What is ANSI?" Joe answered him, "ANSI stands for American National Standards Institute and we can meet those specifications in the engineering drawings." Earl was still concerned. "Has anything come back on the market feasibility study?" "No," Joe pondered, "we need to have a meeting with those people." "Joe, I can't make a trip down there right now. We are in the middle of calving." "Okay, Earl. Let's plan a conference call at our local copy shop. They have a two-way television that we can meet you on. I know you have one up there. Bring your banker and I'll get the market group in with us. We can get some of this worked out," Joe suggested. "Next Tuesday work?" "Sounds good," Earl replied.

The pressure is on the team. The one advantage the A Team does have is their availability to rapid product development tools and the 3D solid modeling. They were able to have a physical part to test the seal very early after the CAD model was built. This speeds up the process by almost 50 percent than the traditional way. The team has already been able to test the part "virtually" and physically. And because all the drawings were digital, Carlos was able to e-mail the pump design drawings to the pattern shop so the mold for the casting could be formed. This allowed the team to use a pattern shop on the other side of the state, which helped keep their costs down since they submitted a low bid. Developing the pattern drawings also went very quickly from the solid model. It just involved making a copy from the model and adjusting dimensions and tolerances.

Carlos was glued to the computer screen as Russell returned from lunch. "Carlos, what are you so serious about?" "Oh, Russell, I just can't seem to make this pump assembly seem clear. I'm afraid whoever is going to assemble this puppy will not be able to distinguish between those two surfaces. They are just too similar and someone could reverse the way it should go together." "Well, why didn't you say that sooner? I have just the solution for you. Move over rover and let me show you how to drive this software." Russell was able to show Carlos how to render lifelike surface appearances and vary the light reflection on the two surfaces so it was now real easy to tell the two surfaces apart. "Well, Mr. Russell! That is real sweet. You're alright! Thanks, that really solves my problem." Carlos exclaimed. Russell retorted sarcastically, "Now you do know how to generate a bill of materials from your CAD database?" "Yes, Russell, I do!" "Remember the old days and all that hand lettering and researching part numbers for parts that don't exist anymore. Boy, I sure don't miss those days. The days of Enterprise Resources Management are here to stay," Carlos continued. Russell added, "It is pretty amazing how right through your computer you can find out how much a

Chapter 6 | Factors That Are Changing the Design Process

A-Team Scenario
continued

part is, what style it comes in, and better yet, import it into your drawing. In the old days, I used to burn so much gas driving around picking up parts or part catalogs." "Guess I better get to work here. I have a conference call with the printer folks for our marketing materials in an hour. See ya!"

The A-Team Design Group are busy as bees: E-mailing, teleconferencing, attaching drawing files and pushing the rapid product development envelope to get Earl's pump to market on time.

Respond to the A-Team scenario by answering the following questions. Use any tools or ideas you have learned in this chapter.

1. Identify ways the A-Team Design Group is utilizing current technology to get Earl's pump to market first.
   _____
   _____
   _____
   _____
   _____
   _____

2. Are there any other technologies or processes that you could identify that could assist the A-Team?
   _____
   _____
   _____
   _____
   _____
   _____

3. Discuss Earl's role in the design process and how it impacts the results.
   _____
   _____
   _____
   _____
   _____
   _____
   _____

TERMINOLOGY & DEFINITIONS

American Society for Quality (ASQ) — key organization for the development of quality systems and education

design for disassembly (DFD) — technique of designing products so that after the product life has been expended, the product can be easily taken apart and the materials can be recycled

rapid prototyping — a digital process that facilitates the quick development of prototypes through a process using 3D solid modeling CAD software and mathematical data. This data is transferred to a prototyping machine which creates a product in wax or resin.

Activities and Projects

- **Review** your design project with your team members. List any of the factors covered in this chapter and how it can affect the final design of your project. Write a reflection paper, together with the team or individually, on the impact the factors can have on your design process and final product.

- **Design** for recycling. Take a common everyday product such as a telephone, refrigerator, or car battery. Research the materials used, the ability to recycle them, and the ability to disassemble them. Write an analytical report on your findings and develop a plan for the re-engineering of the product for DFD/E (Design for Disassembly/Environment).

- **Review** the Post-it® Notes in the chapter that list implications for the designer. Perform a personal review of where you feel you are at with these competencies. In addition, do this with the team as a whole. Develop a chart that shows the strengths and weaknesses of your team. Determine with your team members if the team is balanced in terms of needs for designing your project or if you need to bring in resources or get further training.

- **Develop** a task diagram for your project. Indicate who is responsible for each task and give an estimated time to accomplish the task. This activity should occur at the beginning of the project and intermittently throughout the project until completion.

From Concept to Delivery: Managing the Project

CHAPTER 7

> *"The most critical problem to be faced in modern large-scale complex systems is complexity itself. Unnecessary complexity is largely attributable to the bureaucratic approach to system project management."*
>
> Wilton P. Chase, TRW Systems

Introduction

Today Wilton Chase, author of the previous quote, would have been 91 years old. In 1974 he wrote an incredibly progressive book on whole systems thinking and its relevance to design engineering. At the time of his writings, the United States was experiencing the first impact of foreign competition. Cheap imports were flooding the United States market and the value of imports relative to domestic products rose from 14 percent to 38 percent. Twenty-five years later, engineering has made the shift to systems thinking and management. Are we slow to respond? Maybe, but let's look at the evolution of management and factors that pushed upon it to create the change.

History and Development of a Management System

The evidence of the development of a management system was with the introduction of the railroad in the 1850s (see **Figure 7.1**). The need for systematized data collection and communication to the train crews brought about the first carefully defined structure of management in business. As the railroad grew throughout the country, other businesses grew and expanded their market as well as their need for a management system.

Figure 7.1 Management of business started with the railroad

In the 1920s, General Motors introduced the multidivisional organizational model of corporate management. This management structure has been the model for corporate America and engineering until the 1980s, when global competitors began using a very different management and organizational arrangement that threatened American corporate selling power.

Another major factor that has impacted engineering management is the evolution of computers and information software. During the 1980s United States corporations invested major capital into computers and information technology in hopes of increasing productivity. By the 1990s almost every worker had access to a computer and related software. This did increase productivity but it took ten years to be realized. Much of the profit gains could be attributed to downsizing and the decline in the size of companies as a result of the efficiency of the new technology. Even with new technology, however, companies were still struggling to remain competitive. They realized that the companies' basic organizational structure needed to change to be compatible with the new information technology.

The Old Organizational and Management Model

The management model that brought us to the Information Age is hierarchical in structure. The physical image is a pyramid with workers and field staff on the bottom working to the top with middle management. The chief executive is at the very top (see **Figure 7.2**).

Chapter 7 | From Concept to Delivery: Managing the Project

Figure 7.2
Traditional management model

Each employee had assigned tasks and was responsible to the next level of authority. Important information flowed up the chain of command for decisions, with directives flowing back down for implementation. Within the hierarchical structure, there were other hierarchical structures; for example, the documentation department was placed within the CAD department, a system that was built and constrained by inward controls.

This type of structure does not allow for or foster independent thinking. Its heyday was in the late 1950s and early 1960s, when it served well in the production of large volumes of standardized goods. Because of its rigid structure and linear line of command, this management model has not been able to respond quickly to changes in the domestic and global markets. Its response time does not match the capabilities of current information technology tools.

The New Emerging Organizational and Management Model

In an attempt to recover from World War II, a new form of management emerged in Japan. Along with less labor and fewer resources, Toyota developed a new management approach to production. The Toyota story is well-known to American business people today and many recognize this process as lean production. Lean production combines the advantages of craft and mass production.

Design Tools for Engineering Teams

A COMPARISON OF PRODUCTION PROCESSES		
CRAFT PRODUCTION	**MASS PRODUCTION**	**LEAN PRODUCTION**
• Highly specialized workers • Use of hand tools • Crafted to the design specifications of the customer	• Skilled professionals • Design products made by unskilled or semi-skilled workers • Working on expensive machines	• Combines advantages of both craft and mass production while avoiding the high cost of craft and the rigidity of mass with the goal of eliminating all waste

This form of production eliminates the managerial hierarchy of the old model and replaces it with teams of experts as shown in **Figure 7.3**. These teams include design engineers, computer programmers, and factory workers. This model places all the capabilities needed for the process in one collaborative process as opposed to the old model that separated each level of the process. With Japan's new model came the birth of concurrent engineering.

Figure 7.3 New management model: Team of experts

With the new model of collaboration came the concept of kaizen. Kaizen translates to continuous improvement and the ability to respond to change. To achieve kaizen, management utilizes the collective experiences of the team members. To achieve the strong base of cross-trained resources, an open policy exists for information to flow. The old model held information at the top and sent decisions down the chain of command. The kaizen model puts all the information with the team of experts where the decision is based on firsthand knowledge.

No matter how embellished system management becomes with paper analysis, the real decision-making which shapes the character of system and end item designs is achieved via person-to-person contacts. A small, close-knit system design and development team with appropriate talents and exercising respected leadership over specialized design groups is

Chapter 7 | From Concept to Delivery: Managing the Project

still the only really effective way for producing an end product with desired capabilities.

Does this mean system engineering management is unnecessary? No. It means management resides in the direction of creating a 'total environment' which is conducive to the emergence and effective utilization of creative and inventive talents oriented towards achieving a system approach with a minimum of management encumbrances.[1]

Twenty-five years ago, Mr. Chase predicted what has occurred. In this chapter, we will look at the skills and tools needed to be effective managers and leaders within a system that fosters collaborative decision making in the framework of teams.

As we look at the components of management, we must first understand the difference between project management and organizational management. Projects take place within the framework of an organization and are determined by the abilities of the organization. Budgets, resources (people and technology), and constraints are brought to bare upon a project from the organization. Both project management and organizational management utilize people and resources that are planned, executed, and controlled. The main difference between project management and organizational management is that a project has a start and end to its life cycle, and the organization is ongoing. Design development is typically project focused; therefore the emphasis will be placed on project management in this chapter.

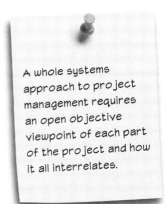

A whole systems approach to project management requires an open objective viewpoint of each part of the project and how it all interrelates.

Industry Scenario

Total Quality Management: A Strategic Paradigm

— *Thomas Misage, CPIM, U. S. Small Business Administration / SCORE Accredited Counselor*

Over the last several years, many companies have begun to realize the need for improving quality in their products and services. Truly successful companies are adopting the total quality management (TQM) strategy to provide the competitive edge in today's market and to set the basis for their future corporate survival.

TQM is a major paradigm shift from conventional approaches to running a business. It requires a directional change and mindset grounded in a *literal obsession* with customer satisfaction. TQM is not a short-term program or a slogan. It involves the focus on the customer by the entire organization in all that is done within the company. Customers ultimately determine the revenue, market share, profit performance, and success of any company. It follows that a focus on customer needs, current and future, is essential to business success. The company must be prepared to delight the customer with highly valued, defect-free products and services for the long term.

Development of a TQM strategy is a difficult and far-reaching effort to understand customer needs and the

---

1 Chase, Wilton P., *Management of System Engineering* (New York City: John Wiley & Sons).

Industry Scenario, continued

company's ability to meet those needs. Top management and company leaders must take the lead in addressing the key elements of TQM to develop a successful strategy. Resolving the following questions will provide the framework for a successful TQM strategy.

How can we:

- understand and define our customer needs and expectations in detail?

- focus the entire organization and all employees on customer satisfaction?

- apply statistical quality control (SQC) techniques throughout the entire organization?

- prevent defects at the source instead of relying on defect detection later in the supply chain or, worse yet, by our customers?

- get the organization to believe that zero defects are possible?

- provide motivation and opportunity to achieve genuine employee involvement and empowerment?

- adopt the paradigm that each employee must satisfy their internal customers and the ultimate external customer in everything that is done and produced?

- measure quality locally at the point of individual accountability?

- assure that company suppliers are involved as an integral part of the TQM strategy and therefore the company's success?

- provide for leadership to drive and implement the TQM strategy?

- make management the models and active champions of the TQM strategy?

- achieve continuous improvement in our processes, products, and in the development of employee capability?

- provide an effective system of recognition and reinforcement for achievements in making the TQM paradigm shift successful?

The role of management in this paradigm shift is critical. They must develop effective and consistent answers to the previous questions at the strategic level. They must provide the needed resources for the implementation efforts. Most importantly, management must endorse and use the TQM strategy and its principles to run every facet of the business. This strategic paradigm starts at the top of the organization and must continually be reinforced in words and actions by leaders throughout the organization. Failure to adopt the TQM strategy results primarily from lack of managerial and leadership support. The entire organization will not embrace the principles of TQM without managerial leadership that is strong and consistent. Management and leaders throughout the organization must act as role models for TQM.

Customers will not support a company that is unable or unwilling to meet their needs and expectations. The progressive organizations have recognized and profited from the reality that superior quality and customer satisfaction are the ultimate competitive weapons in the marketplace.

Stages of a Project

If the focus of an organization is to provide customer satisfaction with their project or product development, then project management provides the road map to get there (see Figure 7.4).

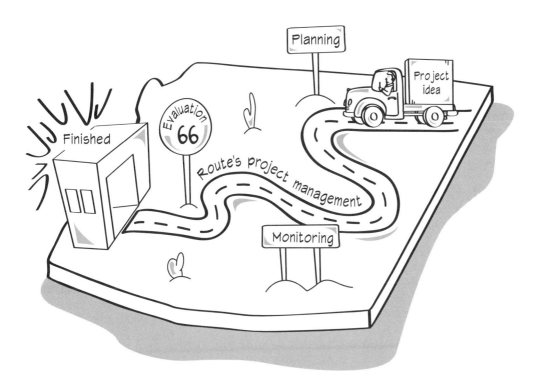

Figure 7.4
Project management is the road map for customer satisfaction

Project management is the plan to accomplish the goal using selected tools and processes to monitor and evaluate the progress. This plan provides a framework for communication and assessment of progress, with a focus on time and cost performances. The tools are methods for tracking progress and the process is communication between different groups or persons involved with the project.

A project has seven distinct stages in its life cycle:

1. Project Definition — The objectives for the project are established. The starting point is determined and a full commitment to the project is made.
2. Project Planning — A game plan for completion of the project is determined. This is based on resources, time lines, and the constraints that would affect the ability to deliver.
3. Project Plan Implementation — This is based on the project plan. People and resources are coordinated to execute the plan.
4. Monitoring Process — A process for measuring the project's progress and a check and balance system for assure that the project objectives are being met.
5. Evaluation and Adjustments — Based on the data produced during the monitoring of the project, evaluation of the data is done with respect to the project's objectives; then, any necessary adjustments are made to meet the project objectives.
6. Closing Process — At this point, the project has been completed and the project objectives have been met.
7. Celebrate — A formal recognition of all participants in the project development process.

    Note: This phase is particularly critical to the life process of a team. Will the team disband or continue together to another project?

Components of Project Management

Depending on the size and scope of a project, there may be many sub-management areas that come to bear on the project. These different subareas become partners with the project team and provide a structured, planned process for the development of a project. Listed here are the different parts of project management.[2]

Project **integration management** includes those activities that ensure that the activities of the project are coordinated. It consists of project plan development and execution and other change control.

Project **scope management** includes those processes and activities that ensure that the project includes all the work required to successfully complete the project. Included areas are initiation, scope planning, scope definition, scope verification, and scope change control.

Project **time management** includes those processes required to complete the project in a timely manner. It consists of activity planning, activity sequencing and duration estimating, and schedule development and control.

Project **cost management** consists of those activities that will ensure that the project is completed within the approved budget. The included processes are resource planning, cost estimating, cost budgeting, and cost control.

Project **quality management** activities ensure that the project satisfies the objectives for which it was undertaken. They include quality processes of planning, assurance, and control.

Project **human resource management** encompasses those activities that ensure the effective use of the people that are working on the project. Included are organizational planning, staff acquisition, and team development.

Project **communication management** is the timely and appropriate generation, collection, dissemination, storage and ultimate disposition of project information. This activity includes communication planning, information distribution, performance reporting, and administrative closure.

Project **risk management** is concerned with identifying, analyzing, and responding to project risk. It consists of risk identification and quantification and risk response development and response control.

Project **procurement management** involves the processes required to acquire the needed goods and services from outside the performing organization necessary for project completion. They include procurement and solicitation planning, solicitation, source selection, contract administration, and contract closeout.

---

[2] Professional Management Institute, *PMBOK® Guide—Project Management Body of Knowledge Guide* (Newtown Square, PA: PMI, 1996), www.pmi.org

Let's look closer at one of these management tools and its relationship to whole systems design.

Risk Management and Systems Design

Risk management is a systems approach to project development. In a risk management approach to systems management, all functions within the process must be described. Each function is assigned a value in terms of its relative importance for meeting the project objective. A function could be equipment, facilities, procedures, personnel or support resources.

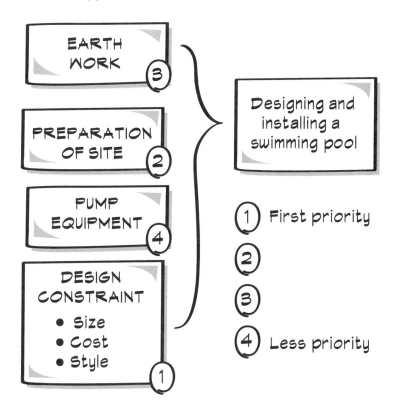

Figure 7.5
Risk management valuates each function of the design process

Overemphasizing or underestimating any particular aspect of the system during a design phase can lead to unnecessary delays as shown in Figure 7.5. Within a large system, there may be multiple subsystems that require planned integration into the whole system. When we compare the construction and design of a house to the multiple systems that integrate into the design and construction of a high-rise building, we can understand how the complexity increases with size. The heating system on a single house involves one furnace unit and duct work which probably will take no more than one or two days to install. The heating system for a high rise is much more complicated in terms of air flow, ductwork routing and overall size of the heating system. The variables increase as the system increases (see Figure 7.6).

Design Tools for Engineering Teams

Figure 7.6
Size of a project determines how many variables there are in the project

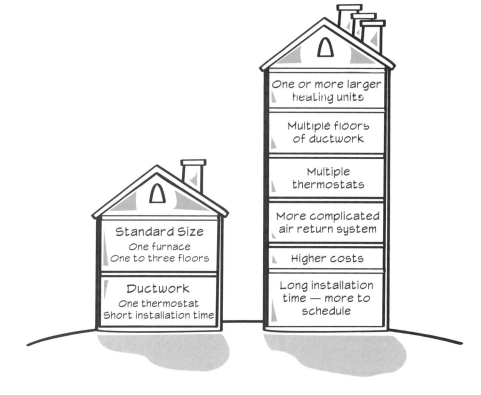

> The team for any project will reflect all aspects of the development of the project.

When we look at risk management, the breakdown of a system into subsystems that include all equipment and services must be related to each function in terms of the role each has in accomplishing the function.

A risk index can be established by quantifying the performance data in relation to the failure data. This is a tool to help determine the weakest links in a system. Once located, development problems can be addressed and controlled.

Planning the Project

Project Constraints/Building the Goals

As the team begins to plan out the project, there are four critical factors that control or constrain a project:

1. Time (the project deadline)
2. Resources (people, money, equipment, technology, information)
3. Specifications/performance (customer quality standard)
4. Unit cost of the finished product

Chapter 7 | From Concept to Delivery: Managing the Project

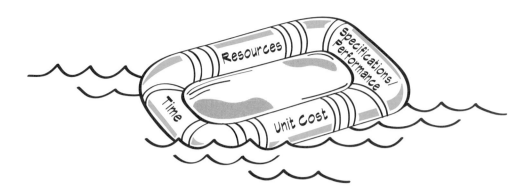

Figure 7.7
Critical factors: Time, resources, specifications/performance, and unit cost

Each of these constraints are like the sides of a rubber raft (see Figure 7.7). Push one down and the others pop up. If the time constraint is pushed out of line, it will affect the cost of the project and may affect the quality. All are related to each other and directly affect the project outcome. It is important to understand the project completely in regards to each constraint before moving forward with implementation. The constraint that will most likely dictate the project will be the one the project originator thinks is the most important. This constraint is important to discover because it becomes the basis of how the project is judged for success. For example, the customer dictated a quick turnaround (time) for a precision valve (quality) and their expectation for cost (money) was low. The product came back with the right specifications and within the time line but costs were high to meet the time constraint placed by the customer. What will the customer accept? Will the customer take the product? Probably not if their most important requirement was cost, but yes if time or quality were their highest priority. So knowing your customer and their priorities in terms of time, cost, and quality dictates the constraints on your project and helps the project team build and set appropriate, realistic goals.

Resources and Time = Cost

Resources for a project can vary from personnel and equipment to computer software. Each project is different and requires analysis of the total resources needed at the early planning phase of the project. It is important to know at this stage the costs and scheduling issues. The management of the project is broken down into tasks and evaluated as to what resources are needed to get the job done, how much it will cost, when it must be completed, and in what order relative to other tasks.

Once resources and time are determined, a budget can be planned out. The budget should account for fixed costs such as personnel salaries, equipment purchases, and so on, but it must also allow for variable costs. All projects incur changes. Fewer changes in the design seem to relate directly to staying within budget since each change increases the costs. That is why it is so important to think through all the possible things that could occur along the way. A team of people helps in this plan-

Design Tools for Engineering Teams

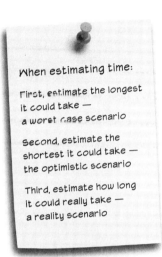

When estimating time:

First, estimate the longest it could take — a worst case scenario

Second, estimate the shortest it could take — the optimistic scenario

Third, estimate how long it could really take — a reality scenario

ning process because each person will give a different viewpoint to make a whole view of the process. The budget now becomes part of the project management plan. It is comprised of labor and material estimates.

Analyze Project Tasks and Flow

Once the goal and objectives have been determined, the next step is to determine all the tasks necessary to accomplish the goal. This could also be viewed as project steps. If too many changes in a project occur without thinking through all the steps, setbacks are sure to occur. Timing and sequencing are critical for the project to move smoothly. A good example is the design and construction of a building (see Figure 7.8). The earth needs to be graded before the foundation can go in, but underground piping for electricity or gas need to go in before grading of the land. There is always a critical path to follow. A critical path is defined as the longest sequence of activities for a project. The steps and constraints along the way should be determined early in the development of a project or there will be mistakes and wasted time or rework when the project development confronts a constraint.

Figure 7.8
Timing and sequence are critical to project flow here.

STEP 1   Evaluate existing grade and conditions

STEP 2   Remove any trees, stumps, or debris where the foundation will be placed

178

Chapter 7 | From Concept to Delivery: Managing the Project

STEP 3   Locate and trench for water and electrical lines

STEP 4   Grade smooth the land where the foundation will be placed

> **Figure 7.8, cont.** Timing and sequence are critical to project flow here.

As we look at integrated engineering, projects involve many different departments and teams to facilitate a project. A plan is essential as well as choosing which tool to use for successful implementation of the plan. There are many management tools and software packages that are available to facilitate this part of the process.

To develop a logical task flow, each task should be charted and evaluated as to the type of task. Tasks fall into three types:

1. Dependent task — a task that cannot begin until earlier tasks are completed
   The foundation could not be installed until the land was graded.
2. Parallel task — a task that can be done at the same time as another task
   Foundation forms for the house can be built at the same time while the garage foundation is drying.
3. Zero impact task — a task that can be done at any time that does not impact other tasks in terms of time or sequence
   Fence can be built around the perimeter of the property of the building project at any time.

Each task should be evaluated in terms of time, sequence, dependency on another task, and other tasks that are independent of each respective task. Considerations should be given to budget, personnel, and assigned responsibility.

Figure 7.9
Task Analysis Chart

Task	What could go wrong	How/When would I know	What can be done
Select software	Vendor software on backorder	When order is placed	Develop contingency drawing schedule

Once the task analysis is completed, a chart is helpful to show the project flow (see Figure 7.9). A chart such as this helps to identify points of constraint.

PERT and CPM

There are two tools that are used together for managing a project. One is the PERT (program evaluation and review technique) method. The PERT method is used to estimate time duration for different tasks. The different time estimates are stated as:

- P = the most pessimistic time
- O = the most optimistic time
- M = the most probable time
- E = time estimate for the task

PERT uses a formula for determining the necessary time estimate for your plan:

$$E = \frac{(O + P + 4M)}{6}$$

Example:

The time estimated to have the foundation completed on the building project is:

- Pessimistic (P) = Three weeks
- Optimistic (O) = One week
- Most Likely (M) = Two weeks

Step 1

$$E = \frac{(1 + 3 + 4 \times 2)}{6}$$

Step 2

$$E = \frac{(4 + 8)}{6}$$

Step 3

$$E = \frac{12}{6}$$

E = Two weeks to complete the foundation

The PERT method was developed in the 1950s and has been used to improve timing from concept to final product.

Chapter 7 | From Concept to Delivery: Managing the Project

The second tool that is employed to maximize utilization of resources for project management is called critical path method (CPM). There are two factors in CPM: Can time be shortened or crashed by putting more resources to it, and can cost be reduced (crashed) by shortening the time constraint?

Example:

Assume each construction worker on a project earns $500 per week in salary.

If one person builds the foundation in three weeks, the cost would be $1,500.

If two people build the foundation in one week, the cost would be $1,000.

The crash cost is $1,000. The crash time is two weeks.

The total savings for this project is $500 and one week of time.

The question is "How much money/resources are saved for what expenditure in time?" The bottom line question is "Was it worth it?" The latter question depends on the project constraints as determined by the project originator.

These two tools can work together to help a team see time and cost savings. By charting out the critical path or paths the sequence of activities for the entire project can be viewed with all the interdependencies as shown in **Figure 7.10**. A critical path is considered the longest path from project inception to completion.

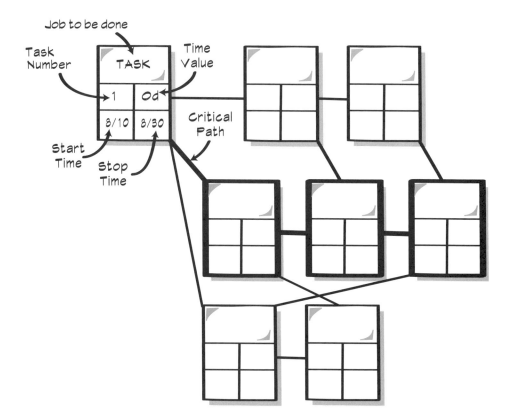

Figure 7.10
PERT and CPM Chart

A quick technique to create a PERT/CPM chart is to utilize Post-it Notes for each task. Arrange the notes in sequence on a white board. Use markers to draw dependencies between each task.

181

Gantt Chart

Another useful tool for project management is the Gantt chart. Gantt Charts graphically show time values and dependencies. It is easy to see the critical path on a Gantt chart. It helps to measure the project against a schedule and determine resource requirements (see **Figure 7.11**).

Figure 7.11
Sample Gantt Chart

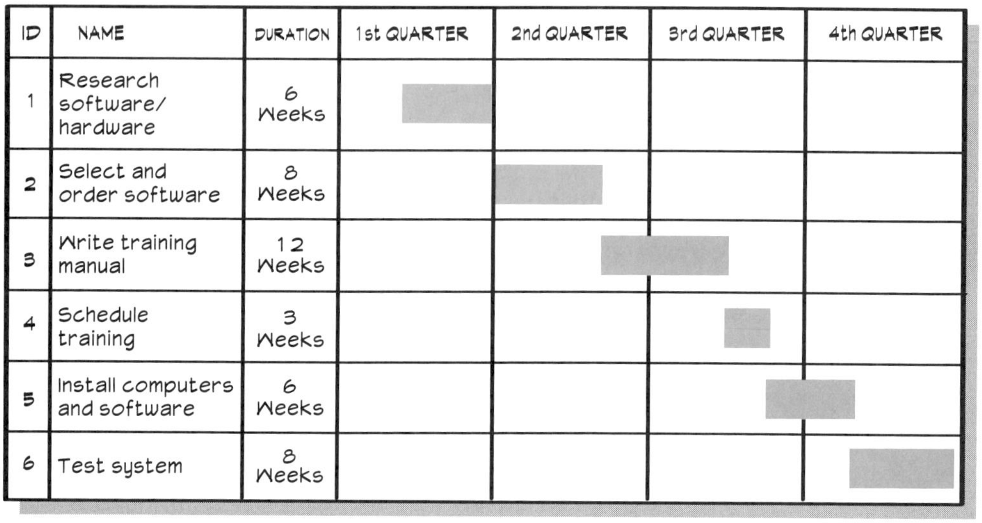

ID	NAME	DURATION	1st QUARTER	2nd QUARTER	3rd QUARTER	4th QUARTER
1	Research software/hardware	6 weeks	▬			
2	Select and order software	8 weeks		▬		
3	Write training manual	12 weeks			▬	
4	Schedule training	3 weeks			▬	
5	Install computers and software	6 weeks				▬
6	Test system	8 weeks				▬

Methods to Shorten the Critical Path

There are three main ways to shorten the critical path of a project:

1. Challenge the time estimate.
2. Challenge the dependency of tasks to each other. Change two dependent activities to two parallel paths.
3. Crash a task — put more resources into the task (overtime/more people).

Understanding the impact of the critical path will help any managing party to better control and monitor the process. It becomes obvious where the bottlenecks are, such as difficult tasks and tight deadlines. As a manager, you can put more resources to this part of the process or shift the workflow to alleviate the potential bottleneck.

Gantt charts, flow charts, and control charts are useful for just a single project or for showing items such as resource allocation (when a piece of equipment is available to do a job). PERT/CPM charts work well to track projects that are interdependent.

Documentation and Specification

It is impossible to discuss project management without also discussing documentation and specifications. A specification is a standard by which the product or design is built to conform. Basically, specifications set guidelines for the end product. The documentation will reflect all the tests, changes, and final documents of a project. The specifications are guidelines for the design and production of the part, plan, or structure. They are used by manufacturing or construction when producing the final product. In the design phase of the project they help define the limits of the design. The management of the design process is driven by the need to meet these specifications.

Specifications can be of various types ranging from the standards set by the customer to the specific industry standards such as OSHA or ANSI. A design specification will determine the production specifications. Specifications make up part of the documentation package, and other documentation for a project could include test results, inspection reports, assembly instructions, and all drawing plans or CAD files.

Management Software

In addition to the basic charts used for management, there are different software programs that are available for project management. They range from very high-end mainframe level to low-end, simple spreadsheets and word processing software. The decision on which to use depends on the size of the project and budget. Today there are excellent PC-based project management programs that are very powerful.

The new software allows the ability to track multiple tasks and the project progress can easily be updated. Another advantage to management software is its accessibility to all team members. Since the data can be kept current, a clear picture of the critical path is maintained so informed decisions can be made relatively quickly.

Industry is finding that using project management software in the early stages of the design cycle helps to save costs and reduce time. As much as 80 percent of product development costs are locked in during the early development stages of the design cycle. Early planning with a software package can allow the multiple users to make course corrections along the way rather than have to rework or reengineer a project or product at the end.

Now with Web-based products available, access for multiple users is also available. Web-based access enables users at all levels to view project information. Product management software can also measure outcomes based on company goals, such as measuring time-to-market savings.

One of the major reasons for project failure is the lack of a structured approach and the disciplined use of management tools. Product management software tools provide the structured approach needed. Now the discipline comes from a commitment from all the team members.

HMMMM... There is Something Holding Up This Project

> H – Human Resources
> M – Materials
> M – Money
> M – Machines
> M – Management

The list above represents all the key components in a project and the possible ways a project can get behind. In each area there are multiple potential opportunities for something to turn out differently than you expected it to. This sounds pessimistic, yet some would say this is realistic thinking. It is important for the team to think through all the "what ifs" and backup options before the project begins. Some call this managing by constraints. Also, do not overlook the simplest things. For example, there will be things that can hold up a project that are out of the control of the manager or team such as whether or not an airline strike affects materials arriving on time. This is when the team will need to be creative and look for alternative solutions until the unexpected holdup gets resolved. So here are a few ideas to consider when planning your project:

- Think of all the ways a project can be delayed or derailed.
- Know all the people involved in the project. Get written commitments for time and cost factors. Inform all parties involved of the performance standards required.
- Communicate the deadlines and keep open communication between all parties. Continuous updates keep everyone on track.
- Utilize information gathered and the expertise gained from other similar projects. Record this project for future information.
- Understand the goal. Send out reminders. It is easy to lose sight of the goal.
- Group together similar operations or activities. This makes it easier to determine needs for HMMMM.

- Remember you are not doing this alone and you are part of a team of experts. Dialog with each other.
- Get management's commitment to be a supporter and protector, not a blocker of decision or micromanager.

The Project Manager

> An effective system engineering manager must be on hand when a problem happens and he or she must deal with it immediately if he or she desires the opportunity to contribute a system-oriented solution. If he or she waits for some system engineering data scheme with its natural built-in time delay to present the problem and a recommended solution, it will simply be a case where all he or she can do is to reflect that, "I wish I had been consulted when that problem occurred." — Wilton P. Chase, 1974

Mr. Chase has said it again. The manager must be down among the design team and ready to respond immediately to any and all issues that arise in the design and development process. Success is likely when the project team has a project manager that serves as their leader.

Management is the process of getting things done through others and a leader must be able to work through other people to accomplish the goal. The shift in project management has come with the shift to teams. The role of the manager has become less autocratic and more like a coach or coordinator as shown in Figure 7.12. On teams, frequently the project manager role switches from member to member as the focus of the project changes.

Figure 7.12
The manager is the coach

The project manager may be appointed by the corporate team or may come from within the team itself. Whatever the advancement to this position the most important truth about successful project management in a team world is to be directly involved with the group (see Figure 7.13).

Managers traditionally have been detached and spent most of their time in ivory tower offices or at the allusive meeting. The truth is summed up by Tom Peters, noted business guru, "The most successful managers spend 75 percent of their time with their people." If the team is going to work for you, you must work with them.

Figure 7.13
The manager is part of the team

One of the other major roles of a team project manager is to be the gatekeeper. It is the job of the team leader to see that the goals are being met and to facilitate team and design reviews. The team leader is responsible for the team's performance. There are times that this means confronting a team member or the whole team to refocus toward the goal.

As the team leader, it is your responsibility to take the initiative to make sure the team continues to work together and more importantly, to see that there is follow through on any action the leader takes.

The most important role a team leader or project manager can have is to instill a vision for the team (see Figure 7.14). This is the difference between a motivated, winning team and a work group. A work group is just doing their job. A team with a vision can achieve great things.

Figure 7.14
A true team leader motivates the team with a vision

The vision comes from both the leader and the team members. The path to reach the goal envisioned comes from the team. As the project manager, you have two choices: you can just solve problems and put out brush fires, or you can build a model of leadership that builds team performance. Your team members are your measure of success.

Improving Performance

A manager would be naive to think that all employees have to do their best work to keep their jobs. Areas like extra effort, high quality, creativity, risk taking, and teamwork are not aspects of performance that are measurable on a human resources evaluation form. The standards usually specify the minimum performance to get the job done. To get employees to meet the challenge of high performance, managers must provide some basics. If an employee is lacking in a skill or ability, then training, mentoring, or coaching could bring this person's performance up to speed (see **Figure 7.15**).

> Team leaders make mistakes too...
>
> A good way to encourage the team to use their mistakes to improve is for the leader to share the mistakes he/she has made and even ask for suggestions from the team.

Figure 7.15
Training for high performance skills

Next, employees must be willing to do what is expected of them. Are they committed to the process? Motivation is key to commitment. Are employee goals in line with the project goals? See **Figure 7.16**.

Figure 7.16
Motivation through goal identification

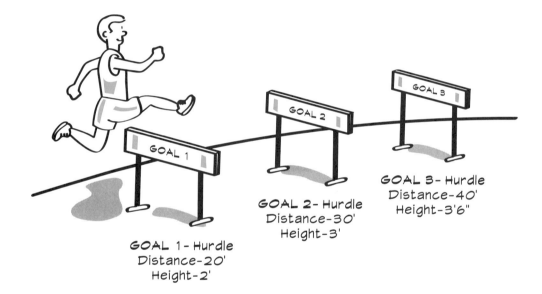

Employees may possess the necessary skills and the commitment to do the job. It is the manager's or team coach's job to support them by setting goals, standards, a clear line of authority and also to provide information, communication tools, and support resources as illustrated in **Figure 7.17**.

Chapter 7 | From Concept to Delivery: Managing the Project

Figure 7.17
Tools to do the job

The quantity and quality of interaction of leaders with their workers has a major impact on performance. Just look at a winning sports team. Not far from the team is a caring, supportive coach that has high standards and expectations for the team's performance.

The two main aspects of performance are employee competencies and abilities, and individual commitment. To improve abilities, the best tool is to diagnose current skill levels and analyze the need for training or mentoring. To improve commitment is to focus on results. The employee is held accountable for achieving the desired outcome of work.

The method to define an individual's abilities is to perform a task analysis. The description of the tasks becomes the basis for determining the individual's level of competence, which helps determine the need for training.

To evaluate performance of the tasks is not as simple. Traditionally this was done through the eyes of the person's supervisor. But as we all know, it's the people we work with all the time that can provide the best reading on how well a worker performs a job.

In the last fifteen years a new tool has emerged to give an evaluation of a person's performance through multiple ratings from people involved with this worker. It is called a multirater assessment (MRA). It is a computer tool that reports ratings of self, peers, superiors, and subordinates and gives a 360-degree evaluation of performance. This "360-degree feedback" is good at identifying, measuring, and improving skills and competencies needed to perform a job. It is also an effective measure of soft skills such as levels of motivation and organizational skills and can be used to identify and measure customer satisfaction, team effectiveness, training needs, and work environment issues as shown in **Figure 7.18**.

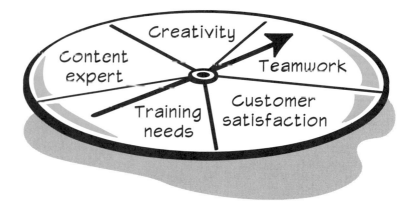

Figure 7.18
360° evaluation gives a person feedback from everyone around him/her

The 360-degree feedback process works to increase and improve performances, but when tied to advancement, it becomes ineffective. The truth and honesty of feedback becomes diminished. Another caution for using 360-degree feedback ratings: this is an automated process and there is nothing better than a good old-fashioned face-to-face discussion. Even though its purpose is specifically to identify strengths and areas for improvement, additional one-on-one communication or feedback may be necessary.

Another important point is that performance results must be linked to organizational goals. One way to tap into a person's motivation is to link incentives with accomplishments. This works for individuals and teams as well.

Summary . . . Links to the Whole

Here is some advice to anyone who finds themselves in a position to manage for continued performance from Deming and Shewhart. These ideas were presented in Japan in the 1950s on the cycle of continuous improvement. The cycle follows four steps.

PDCA or Plan, Do, Check, Act

Step 1:

Plan a change for what you are trying to improve.

Step 2:

Carry out this change on a small scale.

Step 3:

Observe the results.

Step 4:

Study the results and determine what has been learned from the change.

Chapter 7 | From Concept to Delivery: Managing the Project

When change is made, it is made with the idea that it will make things better. If it doesn't make things better, you can learn from the results.

When you improve a process, the corporate knowledge of the process improves as well. Improvement of product and process go together.

For a manager this is important to know and follow. People change with the knowledge that it will improve the process but can only handle small amounts of change at a time.

As we look at managing within a large integrated system, a process for management is critical to product/project success.

Managing the Design from Conception to Final Product or Plan

In this Toolbox you will find many tools, tips, and ideas that can help you and your team better manage your project. Review the tools often so you remind yourself of all the options available to your design team.

- Understanding the stages of managing a project
  - Project definition
  - Project planning
  - Implementation of the project plan
  - Monitoring the process
  - Evaluation of the project/make adjustments
  - Close the project
  - Celebrate
- Management areas of a project
  - Project Integration Management (PIM)
  - Project Scope Management (PSM)
  - Time management
  - Cost management
  - Quality management
  - Human resource management

191

- Communication management
- Risk management
- Procurement management

◉ Knowing the critical factors in a project
- Time
- Resources
- Specifications/performance
- Unit cost

◉ Analyze and identify types of project tasks
- Dependent task — a task that cannot begin until earlier tasks are completed

  The foundation could not be installed until the land was graded.

- Parallel task — a task that can be done at the same time as another task

  Foundation forms for the house can be built at the same time while the garage foundation is drying.

- Zero impact task — a task that can be done at any time that does not impact other tasks in terms of time or sequence

  The fence can be built around the perimeter of the property of the building project at any time.

◉ Three ways to shorten the critical path of a project
1. Challenge the time estimate
2. Challenge the dependency of the tasks to each other. Change two dependent activities to two parallel paths
3. Crash a task or put more resources into the task

◉ Create a PERT/CPM chart to track the critical path of a project

◉ Use Gantt charts to measure the project against a schedule to determine resource requirements

◉ Use project management software if:
- Your projects have more than twenty-five tasks
- Your projects last more than six weeks
- Your projects use more than three resources
- Your projects involve more than $25,000
- Your need to control at least three projects simultaneously
- Your management/clients require regular reports
- You are comfortable using software.[3]

◉ Traits that make a good team coach or coordinator:
- Be involved with and part of the team. Know your people; know their abilities and deficiencies.
- Work with goals and secure commitment from team to their goals. Without goals and the team commitment to these goals, it is like being at sea in a rowboat without oars . . . going nowhere.
- Clarify all expectations to the team. We all need to know what is expected of us.

© 1997, Fred Pryor Seminars, (PM – 0997-US), pg. H8

- Be the point-person for the team. The team coach will need to communicate to upper management and do battle if necessary on behalf of the team, as well as secure necessary resource for the team to meet their goals.
- Value all team members' input. Each team member has their own particular view of the project. Each person's perspective is important.
- Be a good listener. If you are doing the talking, you cannot be hearing the voices of your team.
- Allow the team to disagree with you. An open environment for communication will allow for all issues to come forward that may impact the success or failure of a project.
- Communicate openly and often. Communication from the team leader comes with a trust factor. The leader can do a lot to control rumors, alleviate worries, and get employees back to work faster.
- Develop trust. Trust has to be earned, so always be honest with the team and respect issues of confidentiality.
- Motivate and inspire. If the coach does not root for the team who will? The team will naturally look for the coach's approval. "Positive" is the operative word.
- Be patient. The team leader's attitude that fosters continual improvement through trial and error will allow a team to feel safe to take risks, be innovative, and try new approaches.
- Report all results and celebrate successes. To build team satisfaction, the team needs to know the results of their efforts. If the results were good, celebrate them.
- Be totally committed to the project goals and support their success.

Some common sense rules for conflict and confrontation can help facilitate this process:

- Focus on the goal. Reassert the team goal and isolate the behavior that doesn't contribute to the attainment of the goal. Do not point to individuals — focus on behavior. Always be fair, but firm.
- Remember you are a team. Use words like "we" and "us" rather than "I" and "you."
- Look at confrontation as an opportunity to improve. Focus on opportunities for the team or individual through altered behavior patterns as opposed to criticism that only leads to bad feelings.
- Be specific. Knowing exactly what went wrong allows the team or individual to correct the problem.
- Be positive and honest. Confrontation is never comfortable, but can produce positive results if handled honestly and with a positive attitude for improvement.
- Timing is everything. Waiting for the right moment to confront another may mean the difference between an angry flashback and a person who can hear what is being said to them. Do not be reactive. Wait until the smoke clears, but do not wait too long to confront the other person.

A-Team Scenario

Managing All the Deadlines

The clock is ticking and to the A-Team Design Group it seems to be ticking faster and faster. The pump project is progressing well, but there are a few areas that are holding up progress. One area is the final set of engineering drawings. After the prototype model was generated and tested, changes were sent back to modify the model. With the recent emphasis on the presentation materials, Carlos has not been able to work on the model. Now the pattern shop needs the corrected set as well as the foundry. And the changes also are occurring in the model printouts for the marketing materials. And Carlos is the Master of the Model.

"Oh man, I don't know where to start." Carlos exclaims at the weekly team meeting. I really need your help, guys. I can't do all this by myself." "We understand, Carlos, but not all of us know CAD that well. Claudia responds. Shantel pipes up, "You know I have been reading a lot on project management and some of the software that is now available. Maybe we can get smart about how to manage this project. One of the ways they say to shorten the critical path of a project is to "'crash a task.'" "What does that mean?" Russell asks. Joe explains, "Crash a task means to put more resources on it. Now we could put one more of us on the CAD system and run it round the clock, but none of us are as competent as Carlos. It would take us as long as we have to get the project done, as it will take to get proficient. Or we could clone Carlos; only joking . . . " "Or we could hire a temp from that temp agency in town," Carlos suggests. "I know some CAD jockeys that might want to make some extra bucks." Shantel sighs, "Bucks . . . Can we afford this?" Claudia answers her, "Can we afford not to meet our deadline?" Everyone else speaks out, "Good point!"

The team determined that they would put Russell on the presentation drawings and Carlos would hire a temporary CAD operator for the engineering model updates. They would run on an adjusted schedule so they could triple the CAD time in a day. Shantel went ahead and ordered the project management software. Once it arrived she determined all the jobs that needed to be done and the available resources. The Gantt Chart and the Flow Chart were very helpful to the team to recognize that all aspects of the project were not where they thought. Then Claudia started communicating to the outsource people about due dates. This really seemed to help pull the project back on schedule. Before, the team was really just plodding along and responding to immediate needs. They joked that they were the test case for the one-minute manager. If there is one mistake the A-Team has made, it is not to have a project management tool from the very beginning. They were hopeful it would save the project so they can meet their deadline.

One other lesson the team has learned is that they should have assigned a team leader. Being new at teaming, this didn't occur to them. Actually they tended to look to Joe and Shantel for leadership — Joe because of his depth of experience and Shantel because she is so organized. But they didn't know they were to be a leader because nobody told them, so at times the job didn't get done. The team needs a point-person, someone to keep the goal in front of the group, a person who can see the big picture. Once they realized the project management software was making some positive changes in their way of operating, the team discussed the need to appoint a real team leader.

"I think it should be Joe," said Carlos as they discussed this over an impromptu pizza lunch. "I disagree, Carlos," said Joe. "You all know I'm not gregarious enough. I usually have my head buried in a bunch of calculations. My thoughts are for our leader to be Claudia; she seems to have a good overall sense of the project and also knows all our outsource people." "Excellent choice!" Russell agreed. "Claudia gets my vote," Shantel nods. "She is a lot more patient with things than I am." "Well, Claudia, are you willing to give this a try?" Joe asked. "You really think I'd do OK?" Claudia asked. "Yeah, Claudia, you would do great! Joe is right. You've got my vote, too." Carlos exclaimed. "Well then, I guess I'm it. I think I should say thank you. And you know I will really depend on all of you," Claudia explained.

Chapter 7 | From Concept to Delivery: Managing the Project

A-Team Scenario
continued

Finally the team had formal leadership and Claudia's first job was to get a handle on the results of the project management tool. Her second job was to help everyone know what needs to be done next.

Respond to the A-Team Scenario by answering the following questions. Use any tools or ideas you have learned about in this chapter.

1. Were there any other Project Management tools in this chapter that may have helped the A-Team get on track?

2. Was Claudia the right choice for team leader? Who would you have chosen and why?

3. Based on your chapter reading on team leaders, what must Claudia be aware of to be successful?

TERMINOLOGY & DEFINITIONS

critical path — sequence of activities that represents the longest path to complete the project

cycle time — average time a particular product requires for assembly during the manufacturing process

kaizen — Japanese term for daily continuous improvement

lean production — elimination of waste in the production process through a collaborative method of input and a focus on customer requirements

peer reviews — documented, fully traceable reviews performed by qualified specialists who are independent of the original work, but who have the expertise to perform the work

process control — operating a process, measuring against the requirements, comparing the results of the measurement according to a performance standard, and taking action to correct or improve the process

Activities and Projects

- Using the project you have been working on, determine which are the most important factors for the project.

  1. Time — How long will the project take? How long do you have to complete the project.
  2. Resources — What is the project budget?
  3. Unit Cost — How much will the market pay for the product?
  4. Expectations/Performance — What are the quality standards?

- Write a goal for your project that reflects the priority factors for time, money, and quality.

  _____
  _____
  _____
  _____
  _____

Chapter 7 | From Concept to Delivery: Managing the Project

Activities and Projects (continued)

- Using the following questionnaire, *interview* the client or customer for your project or interview a local company and their customers.

  1. Name of product or project:
  _____

  2. Identify the client's needs in the product/project:
  _____
  _____
  _____
  _____

  3. Brainstorm solutions with the client:
  _____
  _____
  _____
  _____

  4. Describe the client's vision for the end-product/project:
  _____
  _____
  _____
  _____

  5. Identify the client's goals in terms of time, quality, and money:
  _____
  _____
  _____
  _____

  6. Identify the client's resources for help in product/project development:
  _____
  _____
  _____
  _____

  7. Discuss with the client any potential weak links that can be easily identified in the design development process:
  _____
  _____
  _____
  _____

Activities and Projects *continued*

8. Identify computer resources and communication links with the client:

   _____
   _____
   _____
   _____

9. Establish a meeting schedule for design reviews. Bring a calendar to the meeting:

   _____
   _____
   _____
   _____

10. Exchange vital information for future contacts:

    _____
    _____
    _____
    _____

- **Develop** a task analysis chart determining the critical path for a project. You may use the team project or analyze another project in which you have been involved.

   1. First list all the tasks associated with the project.
   2. Identify the dependent tasks.
   3. Identify the sequence for the tasks to be done. Note any parallel tasks.
   4. Determine time factors for each task.
   5. Finally, determine how long the project will take to be accomplished.

Quality through the Design Review Process

CHAPTER 8

> "We cannot direct the wind, but we can adjust the sails."
>
> Unknown

Introduction

In any engineering effort, it is critical to regularly review your progress. Are you still on schedule for completion? Is the product/project still going to do what it was originally intended to do? Does the market/customer still want or need the product/project? Are the costs of development still within the original estimates? If the answer to any of these questions isn't "yes," you want to know as soon as possible. That is why an established design review process is necessary to ensure the success of any engineering effort. The design review is an established component of the engineering design process in which scheduled, periodic evaluations are conducted to ensure that a project or product's quality is being maintained, and that it is achieving the original design criteria.

A design review and analysis should ideally occur at the completion of each phase of the engineering design process. This allows for continuous product improvement, as well as for complete integration of a whole systems approach.

Design review and evaluation activities provide an opportunity to compare the progress of a project against preestablished design criteria. If the design fails to meet the criteria, appropriate analysis and modification activities are implemented to produce an improved product, which is then reevaluated. These activities become part of the continuous quality process until the design goals are achieved. Effective design reviews require that the process be done with a focus on process and/or product improvement. Each predetermined design review phase is tailored to the particular product design, based on project planning and control schedules and design criteria. Formal design review practices reveal errors, omissions, and opportunities in both engineering design theory and application. This allows for the necessary refinement measures and maximum product improvement to occur.

Formal design review processes also focus on other criteria that do not involve specific product design or quality standards, but which have the potential to disrupt or even cause the project to fail. These criteria include:

- Scheduling or time-to-market considerations
- Customer needs and/or desires
- Changing technologies and/or processes
- Management problems
- Product or project maintenance issues
- Reusability and safety issues
- Vendor or supplier problems

The **engineering design review** process typically involves a discussion with all of the project's involved individuals, such as team members, suppliers, customers, and other impacted departments/organizations, in which it is crucial to identify opportunities for improvement, risk reduction, and clarification of ambiguity. The design review is also formally documented and distributed to those who could be affected by the design review changes.

Industry Scenario

— *Sean MacLeod, Business Development Manager, Stratos Product Development Group*

Stratos Product Development Group is a nationally recognized product development firm that uses its combined business and technical skills to design successful products for leading international corporations, as well as aggressive start-up companies. The company's goal of providing leading-edge product development in a wide spectrum of engineering and manufacturing disciplines has led to a reputation for incorporating emerging technologies along with the inclusion of a comprehensive product design and review process.

As Business Development Manager for Stratos, Sean MacLeod's responsibilities involve leading large project teams of customers and design professionals.

"As a former mechanical engineer and now Business Development Manager for Stratos, my main responsibility is to operate within integrated project teams that include representatives from both Stratos and our customer. In addition, I need to readily adapt to our customers' changing needs, and communicate that to our internal design team. I strongly believe in the value of teams, because the reality is that no one individual knows everything. This team approach, combined with persistence, planning, and continuous design reviews, is what I believe leads to success in any type of product development effort."

Chapter 8 | Quality through the Design Review Process

Industry Scenario, continued

"When I speak specifically of the importance of design reviews, there are two aspects that must be addressed. These are the project management aspects of a design, and the technical aspects. The technical aspects include issues such as design for manufacturability and tooling, quality standards, market reviews, pricing considerations, and emerging technologies. The other aspect of the design review that is equally important to address is that of project management. This includes constantly reviewing time schedules, project milestones, and resource allocation issues. A comprehensive approach to the design review process allows for problems to be identified early and solved while they are still manageable.

"In summation, it is important to realize that change is going to happen during the life of a project. There is going to be both internal change and external change. To help minimize the negative impact of these changes, it is necessary to maintain continuous communication with your customers, discussing assumptions, expectations, and defining what criteria will be used to determine success. A major part of achieving this success is through the use of a well-defined design review process."

Components of Analysis and Design Review

Formal documented design reviews should be planned and conducted at appropriate stages of the design process. The review activities should include participants from all functional disciplines concerned with the design stage being reviewed. Other specialist personnel should be included as required. Additionally, customer involvement is also becoming more prevalent throughout the design and build process. Records of design reviews are then maintained as quality records to be used at a later date for reasons such as process improvements, design enhancements, product replication, and customer service. Subsequent to successful design verification, designs will be validated to ensure that the product conforms to quality standards and customer needs and requirements.

The analysis and design review process consists of several components, each of which needs to be evaluated individually and collectively, depending on the project's goals. These components include:

- Product analysis — an in-depth evaluation of the product's strength and weaknesses as compared to similar products developed by competitors.
- Market analysis — an analysis that examines a product's commercial viability in terms of customer requirements and demands, regulatory considerations, competition, and corporate goals/strategy
- Aesthetic analysis — an evaluation of a product's visual qualities to determine if it is pleasing in appearance and/or to the senses. This is commonly part of the market analysis
- Risk management — an analysis of the potential risks involved in the creation of a new product and/or project

- Financial analysis — a study to determine both the cost of producing and the pricing of a proposed product, and its comparison to similar products already on the market or about to be marketed
- Functional analysis — an analysis to determine the proposed product's ability to meet the original design criteria in areas such as performance, cost, safety, and reliability
- Ergonomic analysis — an analysis that measures and validates a product's ability to meet users' physical and safety needs in the most efficient manner

Industry Scenario

A Critical Design Review Case Study: The Mars Surveyor Project

Headed by the Jet Propulsion Laboratory (JPL) in Pasadena, California, NASA's Mars Surveyor Project's primary mission is to complete a total global survey of the planet Mars, over a period of a Martian year (687 days), in a circular orbit. The entire mission will span a total of six years (1996–2003). Upon completion of the survey, the Mars Surveyor satellite is to continue to provide communications for over three years, supplying significant additional survey data. However, upon launch of the Surveyor in November of 1996, damage occurred to the satellite's solar array panels, and it became necessary to re-evaluate the project's mission. The first action taken was to assemble a special critical design review team. The Mars Surveyor Operations Project Critical Design Review was a special review of the Project's planning work by a group of independent technical and management personnel from outside the project.

This review is termed a delta-review (delta, in this case, means an additional increment, or an additional review) because the original team had already reviewed the Mars Global Surveyor (MGS) mission design when it was established several years ago. Other aspects of the project that are reviewed in this manner include the project's requirements, the spacecraft design, the readiness to launch, and the readiness for other major operational events. The delta-Mission Critical Design Review for MGS had the following objectives: first, to validate the strategy and design for the modified MGS mission, and to validate the flight team's readiness for the MGS science phasing orbit. After completion of the review, the group found that the project met the objectives and thus was able to proceed with the implementation of the revised mission design and operations plans. To date, the revised mission plan is on schedule and accomplishing its primary objectives, which include obtaining data regarding Mars' surface features, atmosphere, and magnetic properties.

IMPORTANT NOTE: Because they are designed and implemented by humans, even the best of processes can fail. In October of 1999, NASA's Climate Orbiter, after being lost in space for a week, plowed into the Martian surface and was never heard from again. Due to an embarrassing lapse in communication, engineers had failed to make some simple conversions from English units to metric units. These conversions would have helped Orbiter's position in space relative to moving planets, therefore keeping it on course. Although the errors were subtle, they were sufficient enough to throw the spacecraft off course so that it entered the Martian atmosphere too low and either burned up or was torn apart, causing the complex and costly spacecraft to fail its mission.

Chapter 8 | Quality through the Design Review Process

Upon approval of these initial design review components, the project can begin. The proper sequence and correct implementation of analysis and design reviews are critical to the success of any design project. Successful projects occur when customers get what they want, when they want it, and at a cost they can afford. To accomplish these objectives, a **project plan** is developed prior to the actual development phase. A project plan is a management document describing the specific activities that will occur in a project, including the various design reviews (see **Figure 8.1**). These activities normally include required resources, task breakdowns, evaluation methods, project milestones, quality assurances procedures, timelines, and overall project organization requirements. There are many project management software applications in use today that offer a variety of formats for customized project planning to help develop and manage a project.

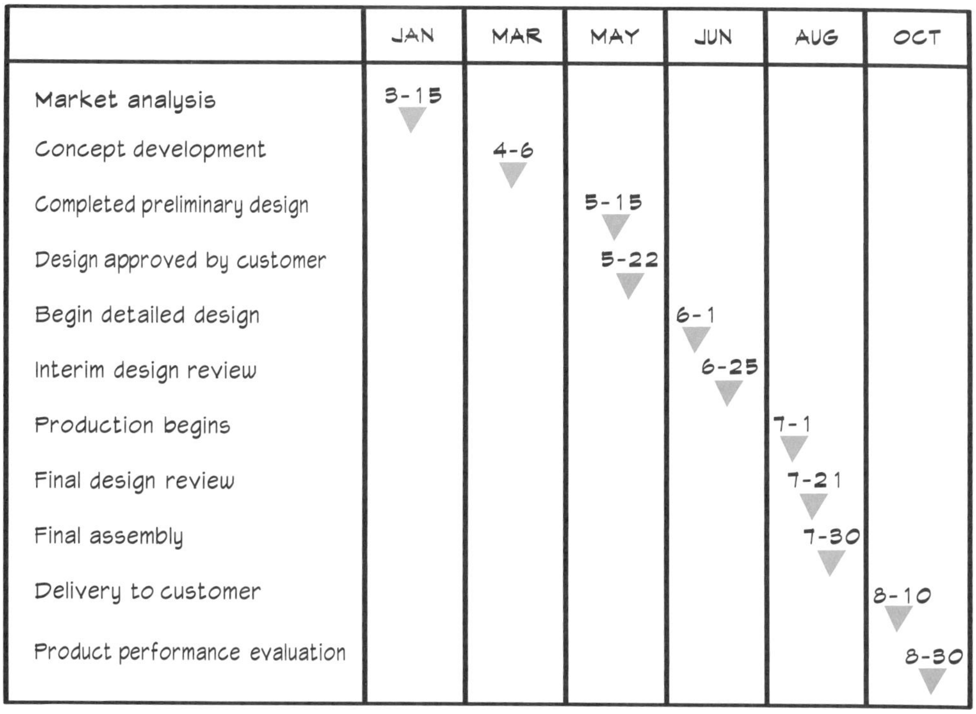

	JAN	MAR	MAY	JUN	AUG	OCT
Market analysis	3-15					
Concept development		4-6				
Completed preliminary design			5-15			
Design approved by customer			5-22			
Begin detailed design				6-1		
Interim design review				6-25		
Production begins					7-1	
Final design review					7-21	
Final assembly					7-30	
Delivery to customer						8-10
Product performance evaluation						8-30

Figure 8.1
Sample of a Project Plan Milestone Chart

Most engineering design review processes contain a checklist as part of the program. The checklist usually contains general questions pertaining to the process, and more specific questions pertaining to the product/project. A typical design review checklist (see **Figure 8.2**) uses yes or no questions to prompt a team response. If the answer to any question is no, then action needs to be taken by the responsible individual or team.

Design Tools for Engineering Teams

Figure 8.2
Design Review Checklist

	DESIGN REVIEW CHECKLIST	YES	NO	N/A
1.	Are design requirements complete and properly documented?			
2.	Are the drawings and specifications complete?			
3.	Does the design solution meet all design requirements?			
4.	Does the design make use of standard parts?			
5.	Does the design comply with good design practice and the design reference handbook?			
6.	Are the objectives of the design review clearly defined and documented?			
7.	Have alternative designs been adequately considered?			

Quality Standards and ISO 9000

The issue of quality is at the forefront of almost all business concerns today. However, the term quality is often tossed around very casually, and there are untold definitions and meanings. But one of the best definitions is that of W. Edwards Deming, the acknowledged master of the quality movement, who defined quality as "exceeding customer expectations through continuous improvement of processes."[1] Quality is the result of standardized processes and a continual search for improving those processes. In engineering, if the design and build process is correctly established and followed, then the resulting product is going to be a direct result of that process. In its broadest sense, *Quality is simply the way we do things*.

This recognition has resulted in the concept of "quality systems." These systems consist of a documented process for achieving quality through the use of formal design and development processes, as well as methods of measuring and documenting project status and progress, commonly referred to as metrics. The most recognized and accepted of these quality systems is that of the International

---
[1] Deming, W. Edwards, *Out of the Crisis*, (Cambridge, MA: MIT Press, 1986).

Organization for Standardization (ISO), a nongovernmental organization composed of the national standards bodies of over 120 countries. The ISO facilitates the development of global consensus agreements on international standards. The ISO has created agreed-upon standards for a vast number of topics and items, but the standard for *quality systems* is referred to by its assigned number, the ISO 9000 series.

ISO 9000 is the part of the ISO standards that is primarily concerned with "quality management." In plain language, the standardized definition of "quality" in ISO 9000 refers to all those features of a product (or service) that are required by the customer. "Quality management" is then what the organization does to ensure that its products conform to the customer's requirements.

ISO 9000 is a process, not a product, and its requirements are generic. It ensures customers around the world that there is a consistency in the process used by a compliant manufacturer. ISO 9000 compliance certification means the company has a recognized quality management system in place. It lends prestige to the company in the international marketplace. ISO 9000 also indicates that the company operates with a constant emphasis on continuous improvement. ISO 9000 certification has become so valued that many companies require their vendors and suppliers to be certified as a necessity for doing business.

The ISO 9000 series includes ISO 9001 (quality assurance model and quality systems requirements) and 9004 (quality management and quality systems development guidelines). After being issued by the ISO in 1987, and revised and published again in 1994 and 2000, the 9000 series has become the most widespread quality standard in the world. Tens of thousands of businesses worldwide have implemented ISO 9000, which provides a framework for quality management and quality assurance.

ISO 9001: Quality Systems, the model for quality assurance in design and development, is the standard for assuring product and service quality through established business processes. Some of the components required for implementing ISO 9001 include detailed processes for:

- Quality policy
- Quality objectives
- Management responsibility

- Contract review
- Design control
- Purchasing and supplier control
- Inspection and testing
- Document and data control
- Handling, storage, packaging, preservation, and delivery
- Training
- Servicing

For all of its seeming depth and complexity, the concept of ISO 9000 can be summarized into the four statements shown in **Figure 8.3**.

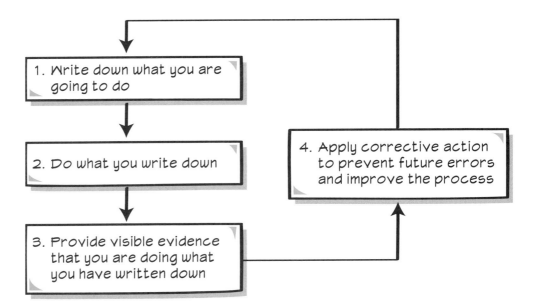

Figure 8.3
The concept of ISO 9000)

For companies to determine what improvements are needed in terms of quality standards, the ISO has established Need Assessment Inventories for each quality component. These self-assessments are an effective tool for establishing a baseline with which to compare later, as well as a real learning experience that speeds up the implementation process (see **Figure 8.4**).

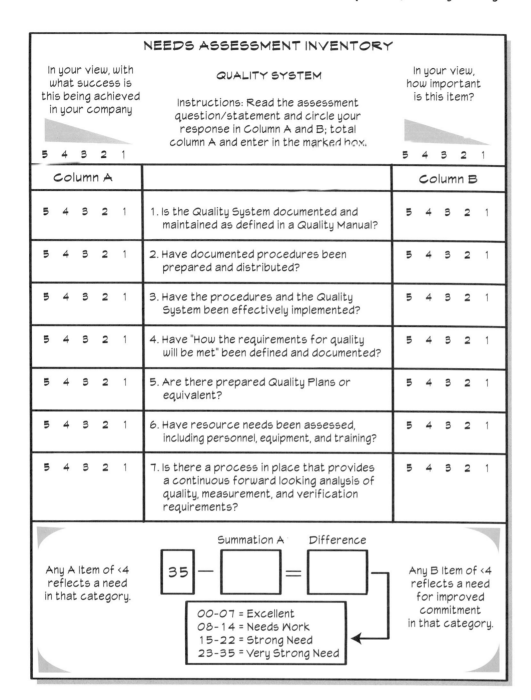

Figure 8.4
Quality System Needs Assessment Inventory

Upon completion of the assessment inventories for each quality component, the ISO has developed a document that allows a company to determine their needs by ranked priority (see Figure 8.5). The information gathered from the inventory can be converted into an action plan relative to the desired quality system.

Figure 8.5 Needs Assessment Listing Ranked by Need

NEEDS ASSESSMENT LISTING RANKED BY NEED

Complete this Needs Assessment by listing the self-assessment numbers contained in the Block marked "Difference" of each section. Then rank in order of magnitude, the largest number (greatest need) first and the smallest number last.

Section		Difference Number	Rank	Comments
4.1	Management Commitment			
4.2	Quality System			
4.3	Contract Review			
4.4	Design Control			
4.5	Document and Data Control			
4.6	Purchasing and Supplier Control			
4.7	Control of Customer Supplied Product			
4.8	Product Identification and Traceability			
4.9	Process Control (Work Control)			
4.10	Inspection and Testing			
4.11	Control of Inspection, Measuring and Test Equipment			
4.12	Inspection and Test Status			
4.13	Control of Non-conforming Product			
4.14	Corrective and Preventive Action			
4.15	Handling, Storage, Packaging, Preservation, and Delivery			
4.16	Control of Quality Records			
4.17	Internal Quality Audits			
4.18	Training			
4.19	Servicing			
4.20	Statistical Techniques			

Even though a vast number of companies around the world have adopted and become certified in the ISO 9000 quality standards, it is important to remember that *people* are the only thing that truly makes quality happen. Although ISO 9000 assures that a quality management system has been implemented and followed, it does not guarantee that the product produced will be of high quality, only that it is likely. Ultimately, it is still the people who design, build, and test the product who determine the level of quality.

Along with Quality Systems, another of the ways companies are integrating a quality approach in day-to-day operations is through the use of teaming (see Chapter 3). Traditionally, mechanics, assemblers, or construction workers would perform a task, and then a quality assurance inspector would inspect it to make sure everything met specifications. If there was a problem, and it wasn't the fault of the worker, it was tossed back to the engineering organization to solve. Additionally, everyone worked in different parts of the factory or worksite. Today, production workers, inspectors, and designers work together to address problems that arise during the assembly and build process. They are often located right next to each other, which, in addition to saving time, improves communication and facilitates an environment where problems can be solved before they occur, producing improved quality and cost savings. Teaming also improves camaraderie among the employees, increases worker knowledge, and allows people to find ways to get things done right the first time. And as business guru Tom Peters often states, "You are either a quality fanatic or you are not in favor of quality at all."

Design Analysis Techniques

Throughout a project's development cycle, the designer turns to the examination of each component to ensure that it is optimized for performance, reliability, and cost. Failure or weakness in any part of the design can often jeopardize the success of the product or project in the marketplace. In addition, design modifications during the later stages of the design process are extremely expensive. As a way to address these concerns, a number of techniques were developed to provide systematic evaluations of specific aspects of the engineering design process. These methods are generally termed design analysis techniques.

A number of design analysis techniques provide, in many circumstances, a cost-effective, rapid, and practical method to meet project goals. There are a number of analysis techniques available to predict such aspects as product performance, life, cost, and reliability, before completed development. These analytical tools are increasingly in the form of software applications that can provide incredibly accurate analysis through the use of computer simulation and presentations.

These tools and their capabilities as part of design analysis and review techniques are used in solving a wide range of design problems. An appreciation of the role that design analysis techniques play in product design and engineering is important. Let's review some of the more common analysis techniques, tools, and methods used:

- Design optimization — a systematic approach to designing a product for the optimal performance, in terms of function, assembly, maintenance, ergonomics, and utilization
- Design for tooling — an analysis tool that considers the engineering costs, in terms of money and design, for any tooling that will be required for the product's development and production

Design Tools for Engineering Teams

- Design to cost — a method that utilizes cost as a design criteria, to help ensure that profit goals are met, and to help control project development costs. Initial cost estimates are derived from throughout the company, including marketing, sales, engineering, production, and finance
- Design for maintainability and reusability — the process for reviewing a product's ability to be reused in a different application and its ability to be maintained after the sale.
- Design for manufacturability — a method for considering the production processes required for a design.
- Finite element analysis (FEA) — an analytical tool used in the engineering of solid materials to determine the static and dynamic responses to specific materials under a variety of conditions.
- Reliability analysis — an analytical tool used for evaluating a product's reliability, in terms of use and/or time (see Figure 8.6).
- Design for safety — an ingredient in the design process that is necessary to ensure that all safety considerations are met in the final product. Designing safety into a product up front not only helps to create a better solution, but prevents potentially costly redesign and liability concerns later on.
- Fatigue damage analysis and life prediction — usually in the form of a computer simulation, this analytical tool allows for formal estimation of a product's functional life span, which is crucial in terms of sales and marketing, pricing determination, and product warranty.

Figure 8.6
Reliability Analysis Process Flowchart

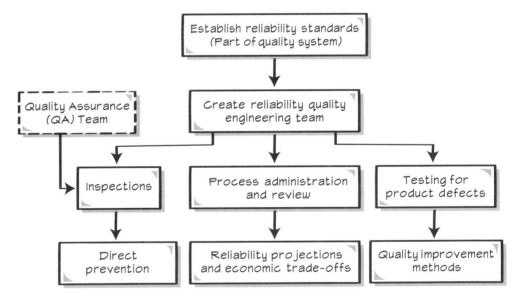

The primary purpose for the use of design analysis methods and tools is to help ensure a quality product or project. At the same time, these methods help reduce the overall costs associated with product/project revisions, errors, and rework.

Chapter 8 | Quality through the Design Review Process

Peer Review Process

Another type of engineering design review is the peer review process. During each phase of the design cycle, when a design is introduced or changed, a design peer review is often conducted to ensure that the design or design change will satisfy the engineering requirements as specified by the design team and/or customer. The peer review process flow is illustrated in the process flow diagram in Figure 8.7.

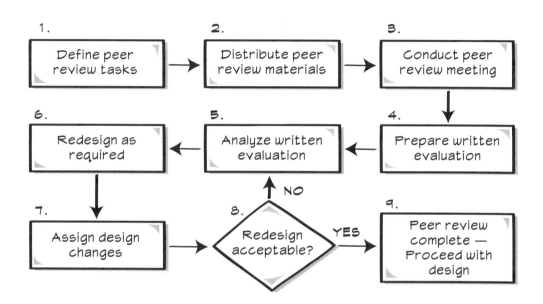

Figure 8.7
Peer Review Process Flow

This process encourages new solutions, suggestions for improvement, or advice from colleagues who have run up against similar design issues.

Industry Scenario

Not Making Assumptions: The Hubble Space Telescope Experience

In 1990, there occurred a mistake that is now used as the classic example of engineering gone awry. All you have do is mention the words "Hubble Space Telescope" and most people will not think of the incredibly valuable information it has gathered about the evolution of our universe, or the breathtaking images it has provided to us, but instead, smile and shake their heads knowingly. Most people still think of the $1 billion error that occurred because the Hubble's 94-inch, 1820-pound primary mirror, the most polished and precise optical instrument ever manufactured, was cut in the wrong shape, or in precise terms, the mirror had a spherical aberration. And furthermore, this error was not discovered until it had been launched into space, put into orbit, and had sent back its first pictures from high above Earth. The pictures were all blurry, as if someone who had never held a camera before had taken them with a cheap Instamatic. How could this have happened?

In selecting a manufacturer for the Hubble Space Telescope's (HST) mirror, NASA selected the prestigious firm of Perkin-Elmer, a premier leader in optics engineering and development. After designing and

211

Industry Scenario, continued

manufacturing the almost 8-foot diameter primary mirror with unprecedented accuracy and smoothness, it was time to test the mirror's curvature. If the curvature was in error by as much as one-fiftieth the thickness of a human hair, the telescope would not function properly. At first, a state-of-the-art measuring device called a *reflective null corrector* was used, and indicated the mirror was exactly precise. Then it was checked again with a less sophisticated instrument called a *refractive null corrector*, which indicated the mirror's curvature was incorrect. In a thorough design review process, there would have been a third test performed, or perhaps the original test instruments would have been checked for their accuracy. But in this case, it was decided that the new state-of-the-art device must be correct, and the older device was wrong, which was the exact opposite of the truth. So the mirror was approved and passed off to NASA as ready to go (see Figure 8.8).

Three years, vast amounts of money, and several complex and dangerous Space Shuttle missions later to correct the error, the mirror is operating flawlessly. What's the moral to this story? A design review process is only as good as its application and the people who administer it.

Figure 8.8
Hubble Mirror representation

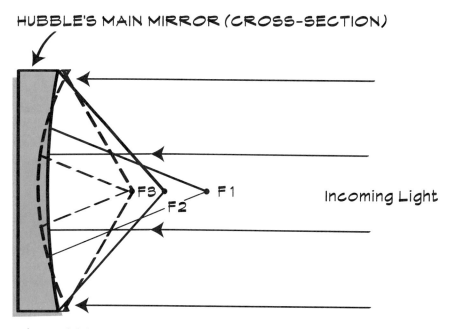

The Hubble Space Telescope main mirror had a defective shape (shaded area) that caused incoming light to focus at different places (F1, F2). A perfect parabolic mirror (original specifications shown by dashed line) would focus all incoming light at one point (F3).

Chapter 8 | Quality through the Design Review Process

Customer Involvement in the Design and Review Process

As with any engineering design concept and process, the ultimate goal is to provide an improved product, process, or service. Historically, projects were developed based upon the "expertise" of the company, with little input from the customer. There was a "throw it over the wall" mentality. This meant that the customer would outline its desires, then metaphorically throw it over the wall to the engineering firm. The engineering firm would then design and develop the project, and then throw it back over the wall to the customer to see how they liked it. Needless to say, customers often received something they hadn't asked for. With the customer-focused business environment of today, customers have become an integral part of the design and development process, from establishing initial criteria to monitoring and approving design and development activities. Keeping the customer involved during the entire life of the project ensures that there will not be any surprises when the time comes to deliver the finished product, whether in terms of cost, appearance, functionality, or any other aspect.

Summary . . . Links to the Whole

As yet another component of the engineering design process, the design review process is a method for not only evaluating the progress of a project, but a valuable tool for taking a moment in the development cycle to stop, and ask the big questions:

- Is this product still what the customer wants, and what they asked for originally?
- Is the project going to be completed on time?
- Is the project/product being designed and built to the quality standards that have been established?
- Have we learned anything that can help make the product better?
- Are the development costs and unit costs still in line with the pricing of the product?
- Have we learned anything that we can transfer to future products or projects?

Knowledge and the proper utilization of the design review process and design analysis techniques are critical in helping to insure the greatest chance at a successful design. These established methods provide some of the tools for designing and building a product to its maximum potential from the very beginning of the engineering process. They provide checks and balances, as well as stopping points in the design journey to ask important questions, and make sure that everyone is still on the desired path. In the end, these processes are a way to continually evaluate and communicate the status of a project to all of the parties involved. This communication will help prevent dissatisfaction and misunderstanding as a project moves from the conceptual phase to the completion phase, and result in a quality product that everyone can take pride in.

TOOLBOX

Quality through the Design Review Process

For Chapter 8, the tools consist of:

- Product analysis
- Market analysis
- Aesthetic analysis
- Risk management
- Financial analysis
- Functional analysis
- Ergonomic analysis
- Project plan
- ISO 9000 guidelines
- Quality System Needs Assessment Inventory
- Peer Review Process
- Design optimization
- Design for tooling
- Design to cost
- Design for maintainability and reusability
- Fatigue damage analysis and life prediction
- Finite element analysis (FEA)
- Reliability analysis
- Design for safety

A-Team Scenario

The A-Team Puts the Pump to the Test

Now that Claudia has assumed the team leadership role, her innate leadership abilities have really been coming forward. With the help of Shantel, she was able to get the project back on track.

"Do you realize we are almost ready to release this pump?" Carlos starts off at one of the weekly meetings. "Yeah, and it is even a few weeks early from our due date." Shantel informed. "That's good!" said Claudia, "because we all still have work to do." "We do?" the group said to each other. Yep, I've been doing some studying on this leadership role you all gave me and I came across some really good materials on the design review process." "What is all this about a 'design review?'" Russell asked. "Well, it's the process of evaluating the progress of a project. I think we are at an excellent stopping point to do that plus we have a little time left to make any adjustments to our pump," Claudia informed them.

The A-Team rallied around the idea of a design review once they understood how it worked. Claudia suggested they include all the outsource people as well as Earl to participate in this review. The most important outcome of this process for the A-Team will be to confirm what they learned in the design process that will produce a better product. They also learned that design is a systematic process. Some components that Joe asked to have included in the design review were an ergonomic analysis and a product analysis.

Chapter 8 | Quality through the Design Review Process

A-Team Scenario
continued

The team discovered they already had a key piece of information and that was their market analysis. Russell asked to add a functional analysis as well.

Though the team now knows that a project plan should have been developed prior to the actual development of the pump, they felt the pump project could still benefit from a design review. "What is so amazing to me," Shantel announced, "is how closely aligned this review process is with our project management software. We can get these two processes working together the next project. It just about gives me chills." "Oh, Shantel, any kind of chart gives you chills. Now just give me a good digital scanner," Russell responded. "We really are pretty far along on the design review and so far not too many red flags." "The design review is a good way to see if the project met the initial criteria for the pump design," Carlos pipes up. "At least we now know the pump needs to be made in multiple sizes to meet the need for high volume versus low volume flow. I think the design review checklist will really help all of us on the team to follow what we are doing. That's a check on Shantel's chart for doing such an outstanding job on the checklist," Carlos applauded. "Alright, back off you two," Shantel chided. "Have you guys seen Joe or Claudia?" "Yeah. They went with Earl to check out an outfit that can perform a finite element analysis on the pump. That seems to be the last piece of information we need. Hopefully these people can tell us how the pump will perform in extremes of temperature and constant vibration from bumpy roads." Carlos explained. "You know, this review has shown me that we should have had vibration as one of our initial design criteria. Well, better late than never. We will add it to our checklist for the next project."

Respond to the A-Team Scenario by answering the following questions. Use any tools or ideas you have learned about in this chapter.

1. Find the different types of analysis components for a design review mentioned by the A-Team. Find the description of each in the chapter and write them here.

   _____
   _____
   _____
   _____
   _____
   _____

2. Are there any other design review components that you felt would be important for the A-Team to include? Why?

   _____
   _____
   _____
   _____
   _____
   _____

Design Tools for Engineering Teams

A-Team Scenario
continued

3. What are some things the A-Team has learned through a design review?

_____
_____
_____
_____
_____
_____
_____

TERMINOLOGY & DEFINITIONS

design analysis techniques — any number of various methods that provide systematic evaluations of specific aspects of the engineering design process

design optimization — a systematic approach to designing a product for optimal performance, in terms of function, assembly, maintenance, ergonomics, cost, and utilization

engineering design review — a component of the engineering design process in which scheduled, periodic evaluations are conducted to ensure that a project or product is accomplishing the original design criteria

ergonomics — an applied science concerned with the design of objects to satisfy the end user in the most efficient and safe manner of operation

finite element analysis (FEA) — an analytical software tool used in the engineering of solid materials to determine the static and dynamic responses to specific materials under a variety of conditions

functional analysis — a method of analyzing a proposed product's ability to meet its original design criteria in areas such as performance, cost, safety, and reliability

TERMINOLOGY & DEFINITIONS

ISO 9000 standards — a set of documented standards developed by the Organization for International Standards (ISO) that describe a quality assurance model in design, development, and management to be used for assuring product and service quality

metrics — the system of measuring and documenting items using quantitative data

peer reviews — documented, fully traceable reviews performed by qualified specialists who are independent of the original work, but who have the expertise to perform the work

project milestones — designated stages in the engineering process where major segments of work are completed

project plan — a management document describing the specific activities that will occur in a project, and when they will occur

risk management — an analysis method used to determine the potential risks involved in the creation/development of a new product, project, or process

Activities and Projects

- Select an ordinary consumer product and perform a product analysis on it. List the product's strengths and weaknesses as compared to similar products on the market.

- Using a product or project related to your specific design discipline (architectural, civil, mechanical, or electrical), perform an ergonomic analysis that considers the product's/project's physical characteristics and usability as required by its users.

- Think about how you view quality. Quality often means something different from person to person. Take a moment to consider the following questions, and then write your answer down on paper:

  1. How do you personally define quality?
  2. Is it something that is important to you?
  3. What role does "quality" play in your education, work, and other areas of your life?

- Select a product that you use in your everyday life. It could be a toothbrush, a car, or a building, whatever you wish. Analyze it for safety considerations. List the positive and negative safety aspects of the product. Write down suggestions of how it could be improved. Be creative!

- Develop a project plan for a project of your choice. Include the required resources needed to complete the project, project milestones, timelines, and design review activities.

Change: Learning to Love It

CHAPTER 9

> "In times of change, learners inherit the earth, while the learned find themselves beautifully equipped to deal with a world that no longer exists."
>
> Eric Hoffer

Introduction

Consider this.... You receive a phone call from one of the most dynamic and successful engineering firms in the country offering you a position with the company. They tell you about the wonderful financial rewards that will be available to you and the opportunities for advancement. But they also tell you that they don't exactly know what these opportunities will be, nor what skills it might take to get them. Nor can they be certain that the company will even be doing the same type of work in the future. The work will definitely be challenging, innovative, and rewarding, but there can be no guarantees regarding the final outcome. Would you accept an offer like that? Well, the fact is that it is the one and only offer available in today's world. We shape the future one day at a time with no road maps, guidebooks, or directions, and constant and never-ending change are the only certainties.

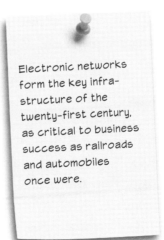

Electronic networks form the key infrastructure of the twenty-first century, as critical to business success as railroads and automobiles once were.

Is it a change revolution? Wait a minute, where has all this talk of change come from? Change has always been around. What's so special about it now? Well, what's so special about it now is the rate at which change occurs (see **Figure 9.1**). Technology has changed the way our world works.

An electronic revolution has given way to an information revolution, a network revolution, and knowledge revolution. Think about it for a moment. Cell phones, pagers, e-mail, laptop computers, virtual conferencing, bar code scanners, global positioning systems (GPS), the Internet, intranets, extranets, tiny silicon chips in everything we can imagine, all connected. In other words, we are talking about global telecommunications and the movement of information in real time. Connecting all of these technologies, while making them smaller and cheaper, has had the effect of shrinking the world.

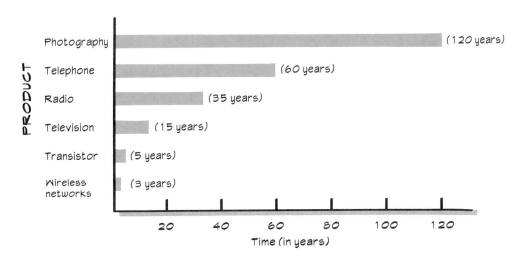

Figure 9.1
Speed of product development from concept to commercialization

We can now work from our home for a company in another city, another state, or even another country. Information is available at our fingertips, up-to-date, and available to anyone else in the world at the same time. The combination of broadband information cables and wireless high-speed data transfers have thrown a net of connectedness over our homes, our offices, and everywhere in between. The world is one huge network! And we are not just talking about the impact on humans either. A rancher in Nevada now knows where all of his cattle are at any given time because of a silicon chip embedded underneath their skin. The Washington State Department of Fisheries tracks the health and location of endangered salmon runs by monitoring the electronic tags that were placed inside the fish when they were spawned at the hatcheries. Almost anything a person can think of, animate or inanimate, can be connected to the information web.

What does this mean with respect to the individual? In terms of the world of work, it means unprecedented challenges and opportunities. The challenges include continuous lifelong learning, shifting careers, and a knowledge of how the landscape of the work world is evolving. The opportunities include the ability to control your career, learn new skills, and find an area of expertise for which you truly have a passion. In the engineering profession, these new realities apply as well. CAD/CAM, rapid prototyping, revolutionary building materials, accelerated product development cycles, shorter product life cycles, and pricing and cost pressures

Chapter 9 | Change: Learning to Love It

are just some of the factors that continue to require the engineering professional to constantly learn, adapt, and anticipate the market for products and services.

Today, many people feel that they are dealing with about as much change as they can handle, yet the pace is not slackening. If anything, the rate of change is increasing, driven by developments like the rapid evolution of new technologies and the growth of international trade. These forces are giving consumers more choice and higher expectations, spurring fierce competition in the marketplace, and creating intense pressure to make more efficient use of resources and lower costs. These forces are creating huge opportunities for nimble and efficient organizations — and brutally exposing the weaknesses of those that are flat-footed or inefficient. And this process is affecting all types of organizations, governments as well as industries, in sectors ranging from manufacturing and construction to education and health care. Is the pace of change likely to let up soon? Don't count on it!

The Nature of Change

Look at each new chapter in the unfolding business revolution of the last ten years, from corporate restructuring to acquisition fever, from start-up mania to the explosion of new entrepreneurs. One dynamic links it all: change. It's not that the business environment is changing. Change *is* the business environment. And it's not that every company is undergoing change. Change has overtaken every company. Creating change, managing it, mastering it, and surviving it is the agenda for anyone in business who aims to make a difference.

Most people are uncomfortable with what they don't know. But there are some people who use that uncomfortableness to their advantage. Change leaders operate that way all the time. In a world that is changing with incredible speed, ambiguity is a constant. Ambiguity defines the work of the change leader: not a comfortable balance, but a dynamic tension between opposing forces. Change is all about taking people beyond their comfort zones. Are you a change leader? Change leaders find themselves working simultaneously across the borders of conflicts — and almost always outside their own comfort zones. And, as mentioned earlier, it's not that change hasn't always existed, it is just the incredible speed with which this change occurs in today's world.

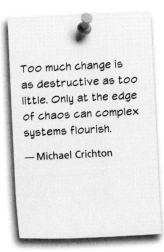

Too much change is as destructive as too little. Only at the edge of chaos can complex systems flourish.

—Michael Crichton

Primarily through the rapid advance of technology, almost any profession or discipline has had to throw out the rulebook. Engineering and computing life cycles (development, production, distribution, and product life) have been compressed into a period of months, not years. An education is now available to almost anyone, anywhere, at anytime, from a huge selection of schools, universities, and private companies through the use of computers, telecommunications, and distance learning models. The world of work is evolving into an environment where a person will switch careers many times throughout their lives. This environment is where a person will be hired to perform a specific skill or complete a specific project, and then move on to the next job, either for the same company, or in many cases, for a new company. The job titles that exist today will not exist tomorrow, however, new ones will. By accepting, preparing, and embracing the reality of change, a person will be able to not only survive, but thrive through their ability to adapt, create, and anticipate the future.

Change is a very messy and complicated business, but it can also be very rewarding. There is an old saying, "May you live in interesting times." These are certainly interesting times. Enjoy them.

Coping with Change

People cope much better with change when they have access to information and are able to participate actively. Rather than becoming a passive victim or active opponent, a person will have to embrace the idea that change is essential, take part in planning the change process, and take ownership of their piece of the action. A person can, and should, become enthusiastic about change.

However, it's human nature to resist change. As creatures of habit, we prefer the security of familiar surroundings, and often don't react well to changes in our environment, even when the changes are positive. This holds true for our workplaces and the security and identity that a steady job provides. Changes in the work environment trigger fears and apprehensions that are deeply rooted in the human psyche. Insufficient information about the impetus for change and its sought-after benefits are likely to cause considerable distress among those affected by the change.

Resistance to change follows some definable patterns (see **Figure 9.2**). One typical reaction is denial, which is perceived as protecting the individual from having to deal with the change. If the change never occurs, there's no need to deal with it.

Passive resistance, another common reaction, often takes the form of anxiety and subtle efforts to slow the pace of change. An individual exhibiting passive resistance will agree on the surface with the need to change, but be unsupportive of it behind the scenes. Active resistance, on the other hand, can show itself as open disagreement with the proposed change. In these cases, people actively lobby against the change taking place and encourage others to do the same.

Recognizing these typical reactions to change, in yourself and in others, will allow you to move through the difficult aspects of change in a more positive manner. It will also allow you to focus on the future benefits of the change event.

Chapter 9 | Change: Learning to Love It

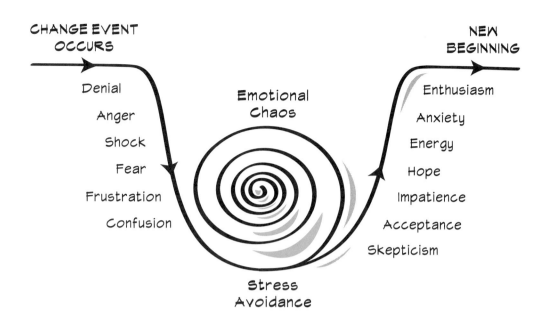

Figure 9.2
Typical reactions to change and transition

The engineering profession is no exception to the rule of constant change — it may even be the leader! The role of technology in society has created an environment in which the engineering world must continually adapt and provide constantly improved solutions and products.

Industry Scenario

The "Education" of Education

For hundreds of years, a formal education was considered a benefit of the privileged. Slowly, civilization has evolved to believe that formal education is necessary for everyone in order to build a truly great society. Although in the past, our culture believed in providing the opportunity for anyone to receive a quality education, the realities of access and cost still effectively limited the choices people had. If individuals were interested in a course of study not offered at local institutions, they had to face the hardships of out-of-state tuition and physical relocation to pursue their dreams. And even if the desired program was offered locally, they were competing for a limited number of openings because the school could accept only so many students a year. But this has all changed, and continues to change rapidly. Through the technologies of computing, telecommunications, and the specific development of the Internet, programs of study from schools, universities, and private institutions around the world are available to almost anyone, anywhere, and at anytime. Education is becoming a commodity, because it is now a widely available service whose content and quality is being driven by consumer demands. This fact has radically changed the way educational institutions operate. They are becoming more responsive to student needs and desires, they are actively promoting their product (education), and they are also learning how to operate in a true market economy. In this new environment, the position of power has shifted from the business (educational institutions) to the consumer (students). This shift in focus is even occurring in elementary and secondary education through the spread of school choice programs, alternative schools, and magnet schools. No one is immune to today's economic and technological realities. Ironically, educational institutions are now being forced to learn a few things themselves.

Types of Change and Responses to Change

Like innovation (as we discussed in Chapter 3), change occurs in one of two ways, either incrementally or radically. Incremental change is that which is constantly occurring as we evaluate and respond to any number of conditions in our lives. Although these changes may be large or small, they represent a continuous process of, we hope, improvement. Incremental change goes on all of the time, as either a *reaction* to some occurrence or thought: "I can't button my pants anymore, I guess I had better lose some weight." Or change happens in *anticipation* of an occurrence or thought: "I think I will go buy some bigger pants so that I can keep eating." The other type of change that occurs is radical change. This is life-altering, fundamental change that causes people to dramatically shift their beliefs or actions. Graduating college, changes in careers, getting married, having a child, or moving to a new city are all types of radical change.

Interestingly enough, radical change usually occurs only because we perceive our lives to be running smoothly and efficiently. When everything seems to be going well, any change is seen as much greater than it likely is. This is one of the primary reasons that radical organizational change in business is so difficult to achieve. Often, the employee level of comfort and satisfaction make the need for change seem unnecessary or even outright wrong. However, it is the ability of a successful business to anticipate and implement change before it becomes mandatory, because by that time, it is already too late. In many cases, this applies to the individual as well. There is a current belief in business that by the time a company's product becomes successful, it is already obsolete. This is because someone else is already working on improving it. The key then is to obsolete your own products before someone else does. Now that's an example of rapid change!

Individuals and businesses can choose how to respond to change. Some are desirable responses and some are not. Let's identify some responses to change by taking a moment to imagine the following scenario: You and a competitor are involved in the finals of the high jump competition for a spot on the school's track and field team. As the competition begins, it begins to rain lightly. You anticipate that the artificial turf you are running on as you approach the high jump will become slick, and so you change to a different type of shoe that holds better on slippery surfaces. In anticipating this change, you have incrementally *fine-tuned* your approach. Your competitor, seeing your change of strategy and its subsequent benefit, also switches shoes. The competitor is *adapting* to the changing situation, but in a reactive manner, not an anticipatory manner. At this point, you become aware that the rain is starting to fall more steadily, the skies are darkening, and a faint sound of thunder echoes in the distance. Rather than make your final jump, you decide to gather up your stuff and head for the shelter underneath the stadium. You have redirected your efforts toward self-preservation as a result of anticipating a radical shift in the conditions. Meanwhile, your competitor is suddenly deluged by driving rain, high winds, and lightning strikes. The competitor is in an *overhaul* situation, radically responding to severe conditions.

As you can see from this story, the ability and necessity for change can be positive or negative. Anticipated change, in the instance of fine-tuning or redirecting, is a position of control and deliberate movement. Reactive change, in the instance of adapting and overhauling, is unplanned and done to maintain the status quo, or even worse, to simply survive (see **Figure 9.3**).

	Incremental change	Radical change
Anticipated	Fine-tuning	Redirecting
Reactive	Adaptive	Overhaul

Figure 9.3
Change Response Diagram

Reducing Resistance to Change

Reducing resistance to change is crucial to the successful implementation of any new business initiative. Since change frequently induces feelings of insecurity, fear, and anxiety, it only makes sense to try and minimize those feelings among the individuals most affected by the change. Here are some specific ideas for reducing people's resistance to new ideas or change initiatives:

- Perceived advantage — The affected individuals should be able to easily see how and why a proposed change would be beneficial to them.
- Compatibility — The better a new idea or change initiative is seen to fit within the current environment, the easier it will be accepted.
- Cost aspect — If possible, the cost of the new idea or proposed change should be less than what it is replacing.
- Reliability — To gain credibility and acceptance, the change must work the way it has been promised.
- Credibility — Does the person(s) responsible for initiating the change or new idea have the necessary experience, education, or perceived integrity to gain the necessary support?

Remember that with any type of new idea or change initiative, regardless of its potential, the perceptions of the individuals affected are more important than the perceptions of the originator.

Managing Change in the Engineering Design Process

Within the engineering profession, the ability to manage change is crucial to project success. Although there are many variations of the engineering change process, the structure and purpose of the process is fairly standard. The objective of any engineering change process is to incorporate design changes as quickly and accurately as possible with a minimum of disruption and cost. Changes in the engineering design process serve many functions: to satisfy customer requests, improve the product, incorporate improvements in production and manufacturing, resolve design problems, or integrate new technologies.

The engineering change process requires thoroughly documented entrance and exit criteria to ensure accurate and timely design changes, as well as a complete history of the changes.

The engineering change process is a major component of the entire engineering life cycle, and the accuracy and efficiency with which it is handled will determine the overall success of the effort.

The Engineering Change Process

As drafters, designers, technicians, and engineers, we would all like to think that after we put the finishing touches on our initial design drawings, we can sit back, fold our arms, and admire our finished work. But the reality is that the work has just begun. These initial drawings probably won't even be recognizable a month from now after they have undergone a design review, various approval processes, and been scrutinized by other organizations involved in the project. Change in an engineering design is as predictable as the sun rising in the east. There are endless reasons why these changes will occur including product improvement ideas, customer requests, engineering drawing errors, or design flaws. The important thing to address is the speed and accuracy necessary to make the changes and communicate them to all of the parties involved.

The engineering change process accomplishes a number of important functions; namely, it serves as a documented and standardized process, it provides a tracking and history of the product, project, or process, and it is part of the continuous improvement process.

The engineering change process can normally be initiated by anyone involved in the project, making it another reason why changes will be inevitable. However, this feature also permits the best possible product, while allowing for potential problems or improvements to occur prior to completion of the product or process. If this type of environment did not exist, failed projects or, at a minimum, costly rework would be the result. The ability of anyone involved in the project to request a change is part of a whole systems approach to the design process.

Let's take a closer look at a typical model of the engineering change process workflow. A tracking document commonly referred to as an engineering change order (ECO), or an engineering change request (ECR), is used in the change process to document any revision that occurs during the design and build process. Figure 9.4 illustrates the engineering change workflow.

Steps in the engineering change process

1. The program generates a unique engineering change (ECO/ECR) number for tracking purposes.
2. Information about the ECO/ECR is gathered from the person receiving the request.
3. The ECO/ECR is assigned to the appropriate person for completion.
4. There is an investigation of the problem and recommendations are made for revision.
5. The actual ECO/ECR is approved to process.
6. The effective date for change is determined.
7. The ECO/ECR is completed by revising existing documents and creating new ones as necessary.
8. The changes go through an approval process.
9. The approved ECO/ECR is sent to all affected parties such as manufacturing, engineering, purchasing, and suppliers.
10. Manufacturing uses the document control process to review and make changes to the designs.
11. Tooling is ordered as needed to manufacture the modified part.
12. The revised part/process is produced/documented.
13. The part/process is incorporated into the project.

Figure 9.4
ECO/ECR Process

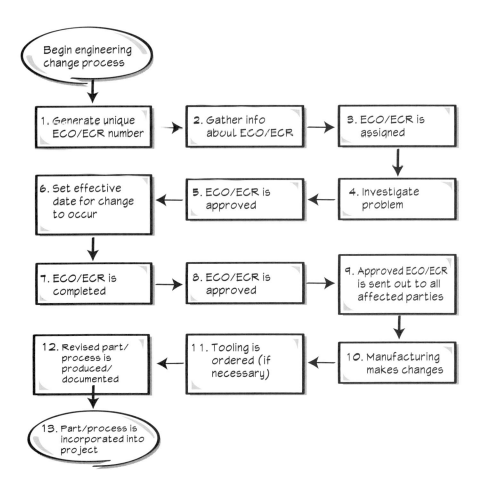

Documents/information required for an ECO/ECR

- Engineering drawings
- Parts list or bill of materials
- Production specifications
- Process specifications
- ECO/ECR document — problem description
- Engineering change order number
- Status of engineering change
- Effective date for change
- Approval list for reviewing changes

The engineering change process can be summarized into three distinct and necessary steps to properly incorporate a change into the design (see Figure 9.5). These three steps are: (1) communicating the change, (2) documenting the change, and (3) tracking the change.

By communicating the change, it is meant that the change needs to be told to the people involved in actually evaluating, approving, and documenting the proposed change. Once this evaluation and approval process occurs, the change must be documented thoroughly and accurately on all affected documents. These documents could include engineering drawings, design standards, parts lists, process standards, production instructions, and so on, or virtually any document that is directly or indirectly associated with the design. Once the affected documents have been changed, they must be distributed to all parties involved, along with the point in the design cycle the changes will begin to take effect. The revised documents also must be stored as part of the historical records of the project. This storage allows the ability to go back and review when and where any changes took place in case of future problems.

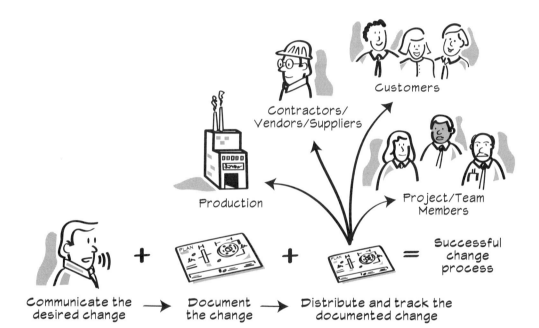

Figure 9.5
Actions to ensure a successful engineering change

Summary... Links to the Whole

Even though we have no control over whether or not change occurs, we do have a choice. We can anticipate and embrace change, and use it to our advantage. Or we can deny that change will happen to us and be regularly buffeted by the winds of change, landing unexpectedly in situations for which we are not prepared. By choosing the first option, we can achieve our personal and professional goals even while the ground around us continues to shift. This can be accomplished by accepting change as a positive force in our lives and careers, remembering that it is change that stimulates motivation, creativity, and growth, both personally and professionally. To successfully navigate the changing societal landscape, we must establish personal beliefs and strategies that will allow us to enjoy life's journey.

Finally, it is important to always view change in the context of the larger picture or the whole system. Remember that change, like anything else, does not occur in a vacuum. What seems to be positive change for some may have negative impacts on others. Therefore, evaluating the effects of change, whether a revision to an engineering drawing, or the decision to not attend a meeting, are critical to making a good decision.

Change... understand it, create it, analyze it, anticipate it, and learn to love it.

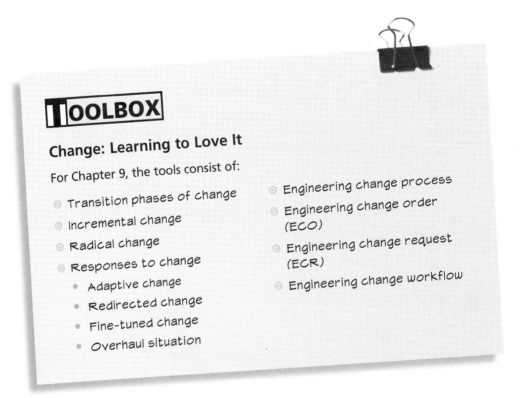

Toolbox

Change: Learning to Love It

For Chapter 9, the tools consist of:

- Transition phases of change
- Incremental change
- Radical change
- Responses to change
  - Adaptive change
  - Redirected change
  - Fine-tuned change
  - Overhaul situation
- Engineering change process
- Engineering change order (ECO)
- Engineering change request (ECR)
- Engineering change workflow

Chapter 9 | Change: Learning to Love It

A-Team Scenario

The A-Team Responds to the Design Changes That Resulted from the Testing

"The results are back from the company that performed the finite element analysis," Joe reported. "Let's go over the numbers." "Boy, Joe, hope you can make sense of all this," Claudia said as she looked over his shoulder. The rest of the team looked on, since they weren't sure what the report meant either. Joe kept reading and started to take some notes. Finally he sat straight up and looked at the group. "Good news. We are okay on the vibration analysis but we have to change the internal bearing so the pump will respond better to the extremes in temperature." "What kind of bearing are we using now?" Russell asked. "It is a standard ball bearing," Carlos reported. "Right, and the FEA guys suggested we use an angular contact roller bearing," Joe replied. "If I remember correctly when looking up bearings on the Web, those bearings are not as cheap as the standard ball bearing," Carlos reported. "Let me get back on the computer and look them up." Carlos was back in a flash with his report. They were more expensive but definitely had the properties to respond to temperature swings. The good news was that the team could easily update the design by importing the new bearing, updating the bill of materials, and adjusting the cost analysis documents all electronically. The bad news was that the cost per pump did increase, but the end result was a better design for the intended application.

CAD/CAM, rapid prototyping, revolutionary materials, accelerated product development cycles, shorter product life cycles, and pricing and cost pressures are just some of the factors that the A-Team has had to address to constantly learn, adapt, and anticipate the market for its particular product.

"Boy, I can't believe how far we have come in the engineering change process," Carlos sighed after completing all the electronic changes. "In the old days that would have taken days to get done. I like this kind of change process." "Carlos, will you coordinate the documentation of all the changes, and communicate them to Earl?" Claudia joined in. "Yes, Team Leader. Will do!"

Respond to the A-Team Scenario by answering the following questions. Use any tools or ideas you have learned about in this chapter.

1. The A-Team experienced minimal changes to the pump design. What would have happened if the pump had to have major changes to accommodate the temperature variances?

   _____
   _____
   _____
   _____

2. What was the A-Team's attitude toward the changes?

   _____
   _____
   _____
   _____

Design Tools for Engineering Teams

A-Team Scenario
continued

3. In what transition stage of change was the group operating?

   _____
   _____
   _____
   _____

4. Did the group effectively address the three necessary steps in the engineering change process (communicate the change, document the change, and track the change)?

   _____
   _____
   _____
   _____
   _____

5. Why is documenting change important?

   _____
   _____
   _____
   _____
   _____
   _____

TERMINOLOGY & DEFINITIONS

change leader — a person who acts as the initiator, facilitator, and motivator of change within an organization or other group environment

engineering change order (ECO) or **engineering change request (ECR)** — a tracking document used in the engineering change process to document any revision that occurs during the design and build process

incremental change — that which is constantly occurring in our lives as we evaluate and respond to any number of conditions in our lives. Although these changes may be large or small, they represent a continuous process of (we hope) improvement.

overhaul change — a radical response to a severe condition that has been thrust upon an individual

Chapter 9 | Change: Learning to Love It

TERMINOLOGY & DEFINITIONS

radical change — life-altering, fundamental change that causes people to dramatically shift their beliefs or actions

redirected change — the type of change that occurs when your efforts turn toward self-preservation as a result of anticipating a radical shift in the conditions

Activities and Projects

- Select a project related to your specific design discipline, (for example, architectural, civil, mechanical, or electrical). Determine an improvement you would like to make to the project and then list the steps necessary to communicate, document, and track the implementation of the improvement.
- As a class, make a list of recent changes in the area of technology. Then discuss whether these changes were incremental or radical in nature, and why.
- In a journal format, write down your experiences in dealing with change. Do you view change as positive or negative? Write about an event that changed your beliefs, desires, values, or approach to life. Was the change anticipatory, or reactive? Why?
- Evaluate a current project you are working on in school. If you suddenly had an idea that would improve the project, whom would you need to tell — teammates, teachers, friends, classmates? Describe in writing how you are going to communicate the change to them.

Delivering the Product

CHAPTER 10

> "Some men give up their designs when they have almost reached the goal; While others, on the contrary, obtain a victory by exerting, at the last moment, more vigorous efforts than ever before."
>
> Herodotus

Introduction

Every project has an end. The focus of any project is to provide the customer with the product(s) and services promised. These products and services are defined within the technical community as deliverables. These deliverables must meet specific and measurable performance characteristics, as well as be within the time frame and budget limitations as agreed to by the provider and the customer. Steps can be taken to enhance the probability that the project will have a successful completion. These include developing well-established acceptance criteria for the deliverables, and up-front planning work to assure customer buy-off of the project, or product, upon delivery. The intended result is a customer who will eagerly return for your services.

Let's take a moment to see what deliverables consist of (see Figure 10.1). By definition, deliverables include all of the documents and components necessary for delivery of your product to the customer, including the product itself. It is important to realize that this is what you need to do to turn all of your work into money!

Deliverables include such items as:

- Engineering and/or construction drawings
- Prototypes
- Life cycle plans
- Installation manuals

- Technical/instruction guides
- Service/maintenance manuals
- Manufacturing process specifications
- Spare/replacement parts
- Product support equipment and materials
- Customer training plans
- Final project report

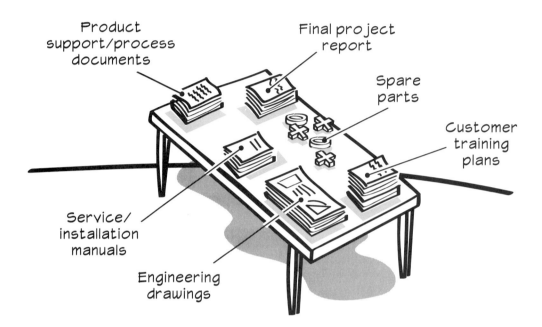

Figure 10.1
Deliverable items to the customer

It is also helpful to think of deliverables in the context of a typical product or project life cycle. As defined by the Project Management Institute (PMI), all work is done to design and produce the project's deliverables. PMI divides any project into five distinct phases. The first four phases of the product/project life cycle are: (1) define, (2) design, (3) develop, and (4) do, which are simply the means to arrive at the fifth and final stage, (5) divest (see Figure 10.2). Deliverables are provided to the customer in the divest phase. By divest, it is meant that the work has been completed, all documentation has been finalized, and final buy-off on the project has been received by the project stakeholders and the customer. So in thinking of all of the aspects discussed in this book about the engineering design process, don't lose sight of one overriding fact: all of the information offered was intended to lead you to the ultimate prize — a product you and your customer can be proud of.

Chapter 10 | Delivering the Product

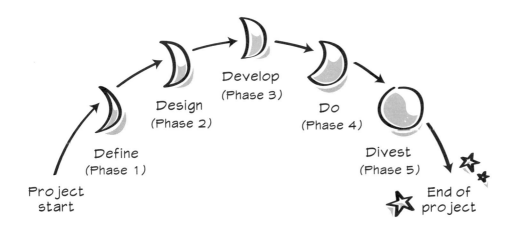

Figure 10.2
Five phases of the project life cycle

Commitments to the Customer

What are your customers really buying when they do business with you? They are quite simply buying a solution to a need, want, or problem. And your credibility and reputation as a business hinges almost entirely on the ability to produce *exactly* what you market, design, and sell.

The company's obligation to the customer is to deliver a product or project that has been built as specified by the initial design requirements, as committed to by the company, and as planned and built to the final specifications. Keeping your commitments to the customer is the company's way of building its reputation, its profitability, and its future existence.

The business principle of relationships states that success in partnering is directly determined by your ability to establish and maintain rapport with your customer. Creating a quality relationship is essential. Your customers will not collaborate with you long-term unless they believe you are acting in their best interests. The higher the level of trust between you and your customers, the lower will be their resistance or fear of making a mistake, and the higher will be their acceptance of your recommendations. But it is important to remember that even if you follow this principle to the letter all of the way through the engineering design cycle, it will be destroyed immediately if you can't deliver the product you promised. And "delivered as promised" means that the product is delivered as requested, as planned, as designed, and as built per specifications. A more valuable goal is to provide a product even better than promised.

Industry Scenario

Amazon.com — Under-Promising and Over-Delivering

It's just about impossible to mention selling on the Internet without the name Amazon.com coming up very early in the discussion. This giant e-commerce retailer of books, music, video, gifts, and more is the model that most other Internet retailers try to emulate. The question then is, even with all of the large retail sellers having established their own Web sites, along with scores of smaller imitators, why has Amazon.com been able to stay the leader of the pack? Well, in the end, their success can be attributed to founder Jeff Bezos, whose philosophy is "under promise and overdeliver" to your customers. Even though customers can get the same books at the same price at a dozen other internet booksellers, people come to Amazon.com because of its reputation for delivering more than the customer expects. Amazon.com delivers their product fast (usually faster than promised), and guarantees the customer's purchase in terms of credit card security, returns, and personal service. They continually offer new services and products, and are always looking ahead to anticipate and meet their customers' desires and expectations. Amazon.com never makes a promise they can't meet or exceed. The company has been able to do this by instilling its values into its employees, and this in turn has given them a significant advantage over their competitors.

The success of Amazon.com boils down to a simple statement: Don't give people what they asked for, give them *more* than they asked for.

Key Elements of Project Completion

If customer satisfaction is the recognized trait of a successful project, then a definition of customer satisfaction must be determined. Customer satisfaction can be defined as the successful fulfillment of all project objectives. Having said this, let's look at some of the key elements in successfully meeting project objectives (see Figure 10.3). Those key elements include (1) time — was the project completed by the date promised to the customer? (2) cost — were the project costs within the initial budget estimates? and (3) performance — does the product perform to the customer specifications?

It's not so important who starts the game, but who finishes it.

— John Wooden

Chapter 10 | Delivering the Product

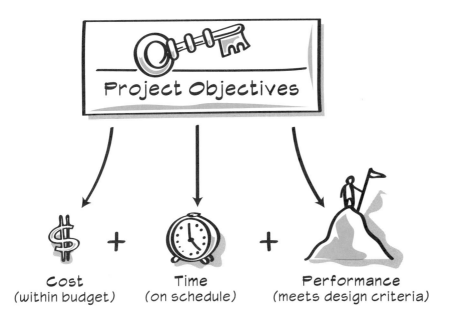

Figure 10.3
Key elements to achieve project objectives

In addition to meeting project objectives, there are other components related to a project that are critical to its outcome. These can be divided into internal factors and external factors. Internal factors are those over which the company has control. These factors are part of every successful company and include items such as a clear and open communication process, well-defined company goals and objectives, management support, priority of project, commitment and involvement by employees, and project and planning controls.

External factors are those over which the company may not have direct control, but must have an awareness of and anticipate the impact of their role in the project. These include such items as knowledgeable customers, regulatory agencies and their requirements, technological innovations, and shifts in the economy and/or customer needs.

These three components — meeting project objectives, addressing internal factors, and accounting for external factors — are keys to insuring a successful project completion.

Completion Activities

There are a number of administrative activities that must be accomplished at the completion of any project. These are activities that should be included in the project's initial detailed project plan. They include (1) a smooth transfer of any necessary operations to customer and support organizations, (2) the reassignment of internal resources, such as labor, equipment, and facilities, (3) documentation, publishing, and archiving of lessons learned from the project including problems encountered, strategies attempted, and solutions, and techniques used that worked well, and also recommendations for future projects, and (4) a ceremony and celebration of your accomplishments!

The last activity, that of holding a meaningful recognition ceremony should not be underestimated. As with the completion of any significant undertaking in life, it is important to recognize the part of your life you spent involved with a project, and to reflect on its value. Taking time to evaluate the positive and negative aspects of a project will permit you to grow and set new priorities both personally and professionally.

Project Completion: Internal Deliverables

When the project is over, it is a valuable exercise to take some time to ask a number of questions before determining the success of the project:

- Have the deliverables met the acceptance criteria established?
- Can the customer use the product or service as desired?
- Did the customer accept the deliverables as given to them?
- Have the "lessons learned" from the project been documented?
- Has the project's historical information been filed for use in future project planning?
- What techniques were not used on this project due to some constraints that might have worked well?

By asking these and other questions, a company will be better able to answer the question of what leads to project success. Also, you will be setting the stage for the next project. The answers to these questions help foster continuous improvement because they enable you to objectively evaluate strengths and weaknesses. And questions will more often than not prompt even more questions, allowing you to discover information you never initially considered. To this end, it is important to develop a requirement for internal deliverables. These are documents or reports that will provide valuable experience, lessons, and insight into future projects. The following is an example of a set of suggested internal deliverables for a technical organization.

PROJECT FINAL REPORT INTERNAL DELIVERABLES

The desired deliverables upon completion of a project include:

- A prototype of the product
- Documentation of the process by which it was developed

This accompanying documentation should include the following:

- Documentation of the problem and the design proposal:
  - a thorough analysis of the problem requirements
  - an extensive exploration of the design process (what are the design alternatives you considered and why did you choose the one you chose)
  - the logic and reasoning used for the proposed design and why it is a good solution to the problem
  - a log of the project meetings to document the previous items
- A discussion of the lessons that one can draw from the project, that is, statements about problems and solutions that would be of interest to all designers, not only to those who are interested in this particular project
- A reflective analysis on the design process (what would the team have done differently if they had to do it again)

Summary ... Links to the Whole

So it all comes down to delivering the goods, both figuratively and literally. In this book, the focus has been to emphasize a holistic approach to the engineering design process. In doing so, many of the skills and abilities that have been discussed are the same ones that can be applied to any personal or professional situation. Almost any project can benefit from looking at a situation from a larger point of view, asking questions, and challenging assumptions. With topics as diverse as creativity and innovation, teamwork, design reviews, research, project management, and the issue of change in today's society, we hope to not only provide information that promotes thinking and questioning, but to illustrate how these various subjects are integral components of the larger process of engineering design. Often, it is easy to become immersed in the incredibly detailed world of the engineering profession. Yet as fulfilling as it is to become a successful member of this rewarding profession, regardless of discipline or position, it is critical to step back periodically and view it from a broader perspective.

As the goal of this book is to provide a larger and more encompassing approach to one's education and work, we should mention that the *deliverable* of any designer or engineer is to contribute to the improvement of our world. It is the engineering professional's job to harness the power of human creativity as it is supported by technology.

The authors wish you all the best in your pursuits, whatever they may be. Strive to remain passionate, questioning, and forever open to new ideas and possibilities.

Design Tools for Engineering Teams

TOOLBOX

Delivering the Product

For Chapter 10, the tools consist of:

- Project/product deliverables
- External deliverables
- Internal deliverables
- Final project report
- Five phases of the project/product life cycle
  1. Define
  2. Design
  3. Develop
  4. Do
  5. Divest
- Three key elements in achieving project objectives
  1. Time
  2. Cost
  3. Performance

A-Team Scenario

The A-Team Presents the Final Pump Design

The big day has finally arrived. The A-Team has been bustling around getting all the documentation packets and pump demo models ready for their debut. Earl arrived early. He was all duded up in his new double-breasted suit. "Do you think we're ready for the big money?" Carlos asked Earl. "Well, from all I can see, your "A-Team" has put together a great package. If they don't finance the production of this pump, they don't know pumps." "Go easy, Earl!" Claudia reprimanded. "Shantel just wants to make sure all the documentation is in the packet so these finance people understand how far we went to design this pump."

The investor group arrived and the meeting went very well. The A-Team design group had all the appropriate paperwork in order from the test results to the final engineering documents. Russell's computer generated three-dimensional animated clips of the pump doing its job of pumping liquid pesticides really impressed them. Shantel's charts and diagrams helped to demonstrate the whole design process, which helped build credibility for the thoroughness of the pump design. At the end of the day, the investor group left with the whole document package and a sample pump. In return, they left a signed contract with Earl for enough business capital to go into full production.

As the door closed behind the investor group, the A-Team and Earl let out a huge hoop and holler. It was time to celebrate a job well done. Earl took everyone out to dinner at his favorite restaurant that served home-style cooking. At the end of the meal he handed each member of the team an envelope. In it was a bonus check for a good job. Earl assured the team he'd be back with some of his other ideas. Riding the range gave him plenty of time to think of these new contraptions.

The A-Team was sad to see Earl leave. He'd been a good partner and easy to work with. It was like they grew up together with this pump design. The team said good night and agreed to meet on Monday morning.

"Good morning, everyone." Shantel greeted everyone. "Well, what do we do now? Our next project won't be in for two weeks." Claudia joined the group. "We're not done yet with the pump." "What!!" "What do you mean?" Everyone was barking at her. "We don't have to make changes do we?" Carlos asked. "No," said Claudia, "but we do have to put this project to bed." "How so?" Russell asked. "Well, for us to improve our designing skills it

A-Team Scenario
continued

is important to review this project for what went well, what we learned, and what we would do differently next time." Shantel chimed in, "I want to put into place in the very beginning of this next project our project management software." "Good idea," Claudia applauded. "We did see the value of that tool in the pump project, didn't we?" "I'd like to see more emphasis on the engineering of the part earlier in the design process," Joe requested. "I agree," said Carlos. "Maybe it would eliminate so many changes in the design."

"One thing that I would like to request is a 360-degree feedback performance review assessment done by all of us and our outsource people," Claudia stated. "Can you explain what that means?" Shantel asked. "Sure, basically it means that all of us give feedback to each other and receive feedback from everyone on our project performance. The intent is to measure and improve our skills and competencies to perform our jobs. It will also identify our team effectiveness and flush out some work environment issues," Claudia explained.

The A-Team continued to complete the review on their pump project and did finally agree to try a 360-degree feedback assessment, even though some were a little nervous about it. They all have learned through the several months of working together on the pump project that honesty and critical review is the only way to improve.

The A-Team Design Group completed their first project. They are definitely a team now.

Respond to the A-Team Scenario by answering the following questions. Use any tools or ideas you learned in this chapter.

1. List some of the lessons learned by the A-Team.

   _____
   _____
   _____
   _____
   _____
   _____
   _____
   _____

2. List some of the positive and negative aspects of 360-degree feedback assessment. Refer to Chapter 7.

   _____
   _____
   _____
   _____
   _____
   _____
   _____
   _____

A-Team Scenario
continued

3. What do you think the A-Team should do differently in the next project?

TERMINOLOGY & DEFINITIONS

customer buy-off — the customer's approval of a specific deliverable

deliverables — all of the documents and components necessary for delivery of your product to the customer.

divest — the fifth and final stage of the Project Management Institute's (PMI) product life cycle. This stage is where the completed work, and all documentation and supporting materials and equipment are delivered for final acceptance by the customer.

"lessons learned" — a business term used to describe the knowledge gained from an event or project which, when properly documented, will provide valuable information in the continuous quality efforts of an organization

project objectives — specific, measurable criteria that need to be achieved to meet requirements for successful project completion

reflective analysis — a written analysis of a design project that focuses on the lessons learned and what might be done differently or better next time

Activities and Projects

- **Select** a project related to your specific design discipline (for example, architectural, civil, mechanical, or electrical). Define and document the deliverables for your project. Additionally, define in writing what a successful project completion will look like.
- **Think back** on the completion of any significant undertaking in your life. Remembering that it is important to reflect on a project's value to you both personally and professionally, take time to evaluate and list the positive and negative aspects of the undertaking you have selected. What did it add or change in your life with regards to personal growth and setting of new priorities?
- **Choose** a school project in which you are currently involved. It does not need to be engineering related. List the project's customers and the internal factors and external factors that need to be addressed to ensure a successful outcome.
- **Select** a recently completed school or personal project. Write down the "lessons learned" from the project including problems encountered, strategies attempted, solutions, techniques used that worked well, and recommendations for future projects.

Bibliography

Rafael Aquayo, *Dr. Deming: The American Who Taught the Japanese About Quality.* A Fireside Book. (New York: Simon & Schuster, 1991) ISBN 0-671-74621-9

John D. Bransford, Barry S. Stein, *The Ideal Problem Solver, 2nd edition.* (New York: W. H. Freeman and Co., 1993) ISBN 0-7167-2204-6

Rosabeth Moss Kanter, John Kao, Fred Wiersema (editors), *Innovation.* (New York: Harper Business, 1997) ISBN 0-88730-771-X

Wilton P. Chase, *Management of System Engineering.* (New York: John Wiley & Sons, 1974) ISBN 0471-14915-2

Kimball Fisher, Steven Rayner, William Belgard, *Tips for Teams.* (New York: McGraw-Hill, Inc., 1995) ISBN 0-07-021167-1

Joseph H. Boyett, Henry P. Conn, *Work Place 2000.* A Plume Book. (New York: Crown Publishers, 1991) ISBN 0-452-26804-4

Frank A. Stasiowski, David Burstein, *Total Quality Project Management for the Design Firm.* (New York: John Wiley & Son, Inc., 1994) ISBN: 0471307874

Marie G. McIntyre, PH.D., *The Management Team Handbook.* (San Francisco: Jossey-Bass Publishers, 1998) ISBN 0-7879-3973-0

Louis W. Joy III, Jo A. Joy, *Frontline Teamwork.* (Burr Ridge, Illinois: Business One Irwin, 1994) ISBN 1-55623-955-6

Jon R. Katzenbach, Douglas K. Smith, *The Wisdom of Teams.* (New York: Harper Business, 1993) ISBN 0-88730-676-4

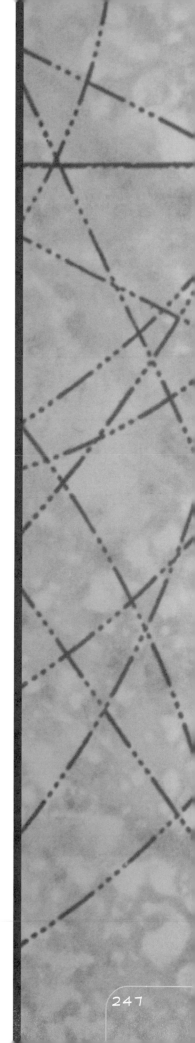

Index

ad hoc teams, 35
    as a strategic planning tool, 114-15

aesthetic analysis, 201

brainstorming, 64-5, 97

causal loop diagrams, 15

change, 219-33
    and product development, 202-21
    coping with, 222-3
    nature of, 221-6
    reducing resistance to, 225-6
    responses to, 224-5
    types of, 224-5

Chase, Wilton P., 167, 185

communication(s), 113-38
    and listening skills, 114, 125-6
    as a strategic planning tool, 114-15
    management, 174
    technologies, 140-2
    with customers and clients, 126-7

computer-aided design (CAD), 67, 142, 147
    and Web sites, 154
    software, 2-3, 150-2

computer-aided engineering (CAE), 67, 142, 147

computer-aided manufacturing (CAM), 67, 142, 147

concurrent engineering, 13, 113-14

cost management, 174

creativity, 55-73
    and the engineering design process, 74-6
    individual versus group, 71
    methods for developing, 61-9
    tools, 67
    versus innovation, 69-71

critical path, 196

critical path method (CPM), 181

customer(s)
    and the whole systems approach, 8
    commitments to the, 237-8
    communication with, 126-7

decision analysis, 94-5

deliverables, 235-6
    internal, 240-1

delivering the product, 235-45

design
    managing the, 167-98

design for disassembly (DFD), 166

design for manufacturability (DFM), 150

design review, 199-204

documentation and specification, 183

dynamic software, 152

e-manufacturing, 145

employee empowerment, 34-5

engineering change order (ECO), 227-8

engineering change process, 226-9

engineering change request (ECR), 227-8

engineering design models, 12-15

engineering design process, 4, 18-23
    and technology, 159-60
    creativity and, 74-6
    factors that are changing, 139-66
    innovation and, 74-7
    managing change in, 226-9
    managing the design in the, 167-98
    teams in the, 31-55
    whole systems and the, 4-9

engineering design review process, 199-204

feedback, 121-2

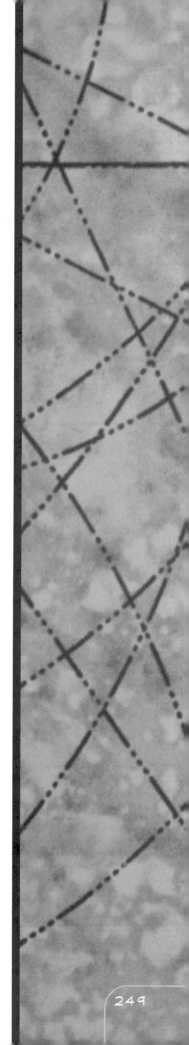

finite element analysis (FEA), 151-2, 210

Gantt chart, 182

group norms, 39-43

holistic approach, 2, 7-9
   definition, 7

human resource management, 174

information sharing, 84-5

innovation, 55-73
   and the engineering design process, 74-6
   and rapid prototyping, 76-7
   versus creativity, 69-71
   through failure, 73-4

integrated product development (IPD), 14

integration management, 174

International Organization for Standardization (ISO) 9000, 204-9

kaizen, 170

kinematics, 152

knowledge-based engineering (KBE), 14

knowledge management systems, 11, 70-1

lateral thinking, 63

lean production, 169-71

life cycle engineering, 13

listening skills, 114
   active, 125-6

management software, 183-4

market analysis, 201-3

Matrix/Checklist Method, 66-7

meeting(s), 123-4

mindmapping, 63, 126

multirater assessment (MRA), 189-90

Murphy's Law, 87, 111

needs assessment inventory, 206-9

negotiation techniques, 127-8

normal group technique (NGT), 98
   organization, 10-11

organizational and management model, 168-9
   new emerging, 169-73

out of tolerance, 87, 111

peer review process, 211-12

performance
   improving, 187-90
   team's, 189-90

Peters, Tom, 186

planning the project, 176-8

presentations, 128-30

problem-solving
   definition, 92
   evolution of, 86-90
   group, 96-8
   information sharing for, 84-5
   process(es), 83-112
   steps in, 91-93, 139
   strategies for teams, 96-9
   to achieve a quality project, 86-91
   TRIZ method for, 67-9

process engineering, 33

procurement management, 174

product analysis, 201

program evaluation and review technique (PERT), 180

project
   completion, 238-41
   planning the, 176-8
   stages of a, 172-3
   tasks, 178-80

project management, 174-5

Index

project manager, 185-7
quality
    through the design review process, 199-218
quality function deployment (QFD), 88
quality management, 174
quality review, 199-217
quality standards,
    and ISO 9000, 204-9
rapid prototyping, 76-7, 155-9, 166, 220-1
recycling, 161
reflective analysis, 241
resources, 177-8
risk management, 174
    and systems design, 175-6
scope management, 174
Smith, Adam, 140
specifications, 183
suppliers, 8
synergy, 37-8
task teams, 35
tasks, 178-80
team(s), 31-55
    charter, 119
    communication, 118-34
    control, 45-6
    cross-functional, 143-4
    definition, 32-3
    development of, 35-7, 39-43
    empowerment, 39-43
    goals, 117-118, 188
    in industry today, 33
    leadership, 38, 45-6
    maturity, 43-4
    meetings, 123-4
    norms, 39-43, 120
    performance, 187-90
    presentations, 128-30
    problem solving, 96-9
    project manager, 185-7
    self-directed, 38
    success of, 37
    synergy in, 37-8, 144
    strengths, 42-3
    types of, 35
    values, 40
    virtual, 144-6
    weaknesses, 42-3
Thomke, Stefan, 76-7
3D modeling, 150-2
360-degree feedback process, 189-90
time management, 174
total quality management (TQM), 14, 171-2
TRIZ problem-solving method, 67-9
variable(s), 87, 111
virtual teams, 144-6
virtual design process, 146-50
visuals aid(s), 128-9
Web-based software, 183
Wheatley, Margaret, 70-1
whole systems approach, 2, 1-30, 167
    and the engineering design process, 4-9
    and competitiveness, 6
    and customers, 8
    and the engineering design process, 4-30
    See also holistic approach
work teams, 35

Notes

Notes

Notes

Notes

Notes

Notes

Notes